NATURAL SCIENCES IN AMERICA

NATURAL SCIENCES IN AMERICA

CLIMATE AND EVOLUTION

By WILLIAM DILLER MATTHEW

ARNO PRESS
A New York Times Company
New York, N. Y. • 1974

Reprint Edition 1974 by Arno Press Inc.

Reprinted from a copy in The American Museum
of Natural History Library

NATURAL SCIENCES IN AMERICA
ISBN for complete set: 0-405-05700-8
See last pages of this volume for titles.

Manufactured in the United States of America

Library of Congress Cataloging in Publication Data

Matthew, William Diller, 1871-1930.
Climate and evolution.

(Natural sciences in America)
Reprint of the 1939 ed. published by the New York Academy of Sciences, New York as v. 1 of its Special publications.
CONTENTS: Colbert, E. H. Foreword.--Gregory, W. K. Preface. Biographical sketch of William Diller Matthew. --Climate and evolution. [etc.]
1. Paleontology. 2. Zoogeography. 3. Bioclimatology. 4. Paleoclimatology. 5. Evolution. I. Title. II. Series. III. Series: New York Academy of Sciences. Special publications, v. 1.
QE721.M3 1974 560 73-17830
ISBN 0-405-05748-2

CLIMATE AND EVOLUTION

Peter Van Valkenburgh

William Diller Matthew 1871-1930

Special Publications of the New York Academy of Sciences

Volume I, pages i-xii, 1-223

Erich Maren Schlaikjer, Editor

CLIMATE AND EVOLUTION

By WILLIAM DILLER MATTHEW

SECOND EDITION, REVISED AND ENLARGED

ARRANGED BY

Edwin Harris Colbert

PREFACE BY

William King Gregory

WITH CRITICAL ADDITIONS BY THE AUTHOR AND OTHERS AND A BIBLIOGRAPHY OF HIS SCIENTIFIC WORKS

by Charles Lewis Camp and Vertress Lawrence VanderHoof

June 15, 1939

Publication made possible through grants from the income of the Permanent Fund and the Ralph Winfred Tower Memorial Fund.

CONTENTS

FOREWORD

Since the publication in 1915, in the Annals of the New York Academy of Sciences, of Dr. Matthew's essay, "Climate and Evolution," there has been a steady demand for this paper that has continued to the present time, long after the original edition had been exhausted. Indeed, this demand has increased to a marked degree during the past few years, and for this reason the Publication Committee of the New York Academy of Sciences has deemed it advisable to issue a new edition of the work.

Accordingly, a committee, consisting of Dr. William King Gregory, Dr. Harold E. Vokes and the writer, was appointed to supervise the preparation of a new printing of "Climate and Evolution," and to decide just what such a new edition should contain. The task of selecting and preparing material for the new volume was entrusted to me.

In the years immediately preceding his death, Dr. Matthew had made many annotations in his personal copy of "Climate and Evolution," presumably with the idea of bringing out a revised edition at some future date. These annotations are included in this new volume. Speaking for the committee, I wish at this place to express our sincere thanks to Professor Charles L. Camp of the University of California, and to the Matthew Memorial Library of that institution, for allowing us to use Dr. Matthew's annotated copy. Our appreciation is due also to Professor Camp for sending us the unpublished manuscript that was found among Dr. Matthew's papers and notes, and for giving us permission to publish it here for the first time. The complete annotated bibliography of Dr. Matthew's published papers, presented at the end of the present volume, is the work of Professor Camp and Dr. V. L. VanderHoof, of the University of California. Our deep appreciation is due them for the interest they have taken in this work, and for the cordial cooperation that we have received from them during the progress of preparing the new volume.

Acknowledgments are due also to Eunice Thomas Miner, who has devoted much time and effort in her capacity as Executive Secretary of the New York Academy of Sciences, to making the publication of this volume possible. Mrs. Miner transcribed Dr. Matthew's annotations, and in many other ways assisted in preparing the volume for the printer.

Finally, our thanks are due to the Geological Society of America and to *Scientia,* for permission to reprint the articles entitled, "The Affinities and Origin of the Antillean Mammals," and "The Dispersal of Land

Animals." Also to the University of California Press for permission to reprint figures 8 to 12 inclusive, from Syllabus number 213.

In the present volume Dr. Matthew's annotations are added to the text in their proper places. They are, however, included within double parentheses. New words or phrases, substituted for the original text, also are placed within double parentheses. New footnotes made by Dr. Matthew are placed within double parentheses, while those made by the writer are bracketed, and followed by the initials, E.H.C.

EDWIN H. COLBERT

American Museum of Natural History, New York
February 7, 1939

PREFACE

By William King Gregory

A BIOGRAPHICAL SKETCH OF WILLIAM DILLER MATTHEW 1871–1930[1]

The death of William Diller Matthew on September 24, 1930, cut short at the height of his career a man who had contributed immensely to the science of paleontology, one who realized the value of this science for the philosophy and art of human living and who was in the midst of a sustained and highly successful effort to open wide its broad fields, especially to great numbers of his own well-trained students.

In 1894 Matthew was a red-cheeked Canadian youth in the department of geology at Columbia. His main interests at that time were in crystallography, in trilobites, and in the structure of the intrusive and effusive rocks of his native country around St. John, New Brunswick. His father, George F. Matthew, was an amateur geologist in the best sense, since he was a recognized authority on the geology, fossil plants, and early amphibian footprints of New Brunswick. His mother successfully reared a large or "old-fashioned" family and imparted to all her children her irrepressible cheer and good humor, as well as her high ideals of service. Fire consumed the elder Matthew's collection of fossils and his scientific library; but he set to work to build them up again. The son was a young man of almost Spartan simplicity of life: frugal and self-denying but early learning to achieve excellent results with slender resources.

With such a background the young Matthew, after obtaining his doctorate at Columbia in 1895, came to the American Museum of Natural History at the invitation of Professor Henry Fairfield Osborn to be a scientific assistant in the then young department of vertebrate paleontology. His first big task was to go down to Philadelphia to catalogue and pack up for shipment the great private collection of vertebrate fossils which had been amassed by Professor E. D. Cope, and had recently been sold to the American Museum. In this way Matthew gained his first extensive contact with Cope's life-work and collection. During the next thirty-five years it was his lot to catalogue and identify tens of thousands of vertebrate fossils for the ever-growing collections of which the Cope collection

1. Excerpted from an article in Science 72 (1878): 642-645. December 26, 1930.

was the foundation. More cautious than Cope and far more critical, with the advantage of great stores of additional material, he corrected, revised and extended Cope's work in many fields, conserving and strengthening as much as he reasonably could of Cope's results. For instance, in all his extensive researches on the classification of the Eocene carnivores he clung tenaciously to Cope's definition of the Creodonta, which had been so framed as to include both the typical specialized creodonts and their differently specialized relatives, the ancestors of the higher carnivores. This was not because he doubted the ancestral kinship of the miacid creodonts to the higher carnivores but simply because, like Cope, he was a stout defender of the "group method" of classification, which insists on the formal definition of groups of related organisms that possess in common the characters cited in the definition, this method being in opposition to the "phylogenetic method," which defines "phyla" according to their supposed trends of evolution and which includes in these phyla even the earliest and little-differentiated representatives that have not yet acquired all the visible characteristics of their descendants.

Another instance of his conservative progressiveness may be found in his most widely known work, "Climate and Evolution" (1915). Being a student of Alfred Russell Wallace's works on geographic distribution, he sided with that great naturalist in opposing those who lightly reconstructed "land-bridges" across the oceans of any and every geologic epoch in order to account for the presence of related animals and plants on opposite shores of ocean barriers. From his vast and intimate knowledge of the fossil record of the vertebrates, he was able to show that many groups whose representatives are now found dispersed in far distant tropical and southern continents are the specialized descendants of known fossil types which in the Eocene and later epochs were inhabitants of the northern lands. His excellent distribution maps were "North Polar projections" of the continental land masses, upon which was plotted: first, the distribution of the Eocene and later ancestors of the horses, tapirs, rhinoceroses, camels, pigs, ruminants, dogs, cats, etc., which were all known only in the northern hemisphere; second, the present distribution of the scattered representatives of many of these families in tropical and southern land masses. His inference was that these forms had originated in the northern hemisphere, had then spread southward into the tropical and southern countries, while the original stock often became extinct in the northern homeland.

On the geologic side he adopted the theory of isostasy, in so far as it was developed at the time. According to this theory, the continents and oceans are held in balance in such a way that only minor oscillations of level have occurred; thus, although the land was at intervals submerged under shallow invasions of the sea, while at other times the continents emerged to higher levels, yet the main continental masses have remained substantially intact, at least throughout the period covered by the records of vertebrate life. In order to account for the first colonization by mammals of continental islands, including Madagascar, Australia, New Zealand, he followed Wallace in evoking the agency of those "natural rafts," which emerge in great numbers from continental rivers and carry sometimes a stray small mammal. Once in a million or more chances, he argued, this castaway might be a gravid female whose progeny could subsequently colonize the whole island. He cited the evidence that tended to support the view that even the most gigantic Australian marsupials had been derived eventually from small arboreal ancestors, which were the only kind that might be conceived to be capable of living in a tangled mass of vegetation during its long drift across the sea and through subsequent perils of landing on a strange shore.

Thus Matthew came to deny the validity not only of all the supposed sunken "land-bridges" across the Atlantic and Pacific which had been evoked by various authors, but also the supposed former connections of South Africa, Australia and South America with the Antarctic continent. Accordingly he ascribed to "parallelism" the rather striking resemblances of the extinct "sparassodonts" of Patagonia with the existing thylacines of Tasmania.

Whatever may be the ultimate verdict of science upon these features of his zoögeographical theory, it is somewhat unfortunate that this more or less debatable aspect of his work should have been discussed by dozens of writers of varying competence, while his immense contributions to the orderly development of mammalian paleontology were really well known only to those two or three specialists who, with fossil specimens in hand, wanted to identify their material by comparison with his excellent descriptions and figures.

The greater part of his scientific writings deals with the revision of mammalian fossil faunae, including the Basal Eocene Puerco and Torrejon, the Lower Eocene Wasatch, the Middle Eocene Bridger formations of the Eocene and many horizons of the Oligocene and later Tertiary. To all these faunae he contributed extensive technical reports or memoirs. His faunal lists of the Tertiary horizons of Western North America, though embodying the work of his predecessors and colleagues, nevertheless represented his own labors in identifying thousands of fossil specimens; they were also based upon his personal studies of the precise geologic level of specimens collected in field parties under his direction.

He was always a geologist as well as a paleontologist and geological considerations occupy a good share of his published writings. As a result of his field and museum experience he successfully attacked (1899, 1901) the theory that the deposits of the supposed ancient "Lake Basins" of the West had been laid down in great lakes, showing that the geological and paleontological facts indicated rather the derivation of these formations by flood-plain playa and aeolian deposition.

His conclusion that the Basal Eocene formations of New Mexico and Wyoming represented a very long period of time (to which he gave the name Paleocene) between the Uppermost Cretaceous and the true Eocene has recently been confirmed by the intensive explorations carried on in northwestern Wyoming by parties from Princeton University.[2]

Owing to the great scarcity of fossil mammals, except in a very few museums, and to the necessarily technical nature of most of his work, Dr. Matthew's greatest discoveries concerning the evolution of the mammals were known at first hand only to his immediate colleagues and to a very few specialists scattered over a world which, for the most part, is impervious to paleontological science. But, as practically his entire scientific career was spent amid the most compelling evidences of evolution, it is not surprising that when he did write for the public on the subject he was able to do so with conviction and with authority. His vigorous honesty and cautious testing both of fact and of inference inspired confidence in the reader. Among the most important of these writings on evolution were his popular "guide" to the evolution of the horse[3] and his admirable articles on the evolution of the horse family and of the dog family,[4] his summary of the evolution of Eocene mammals in the Proceedings of the Zoological Society of London (1928), and especially his handbook "Outline and General Principles of the History of Life."[5]

In 1927 Dr. Matthew accepted the call of the University of California and went there to be head of the department of geology, professor of paleontology and director of the paleontological museum. He was brilliantly successful in attracting very large numbers of students to his lectures in spite of the difficulty of his courses; he also stimulated new exploration and research and attracted young men of great promise into the

2. JEPSEN, G. L.: Proc. Amer. Philos. Soc. **69**: 467, map.
3. Quarterly Review of Biology **1** (2). April, 1926.
4. Journal of Mammalogy **2** (2). May, 1930.
5. University of California Publications, Syllabus Series **213**. 1928.

field of paleontology. In the summer months he came back to his old collections in the American Museum to continue and complete his monograph on the Paleocene mammals. But in the midst of all these successful activities he was interrupted in May of 1930 by the first serious indications of the grave illness to which he finally succumbed after a long fight.

Even in a much fuller review of his work[6] it has been impossible to do more than touch upon a few of the evolutionary problems which he either definitely solved or left with significant enrichment. It must suffice in this place to state that the younger generation of American paleontologists, which is now fortunately coming forward, is already finding that Dr. Matthew, while giving final answers of fact to thousands of specific questions, has also bequeathed to them other thousands of problems that will challenge their best efforts for a lifetime.

[6. William King Gregory: Amer. Mus. Novitates 473. 1931. E.H.C.]

Reprinted From
Annals of the New York Academy of Sciences
Volume XXIV, pages 171-318
February 18, 1915

CLIMATE AND EVOLUTION[1]

By William Diller Matthew

CONTENTS

1. Presented in abstract before the Academy, February 13, 1911. Manuscript received by the Editor October 20, 1914.

THESIS

1. Secular climatic change has been an important factor in the evolution of land vertebrates and the principal known cause of their present distribution.

2. The principal lines of migration in later geological epochs have been radial from Holarctic centers of dispersal.

3. The geographic changes required to explain the present distribution of land vertebrates are not extensive and for the most part do not affect the permanence of the oceans as defined by the continental shelf.

4. The theories of alternations of moist and uniform with arid and zonal climates, as elaborated by Chamberlin, are in exact accord with the course of evolution of land vertebrates, when interpreted with due allowance for the probable gaps in the record.

5. The numerous hypothetical land bridges in temperate tropical and southern regions, connecting continents now separated by deep oceans, which have been advocated by various authors, are improbable and unnecessary to explain geographic distribution. On the contrary, the known facts point distinctly to a general permanency of continental outlines during the later epochs of geologic time, provided that due allowance be made for the known or probable gaps in our knowledge.

INTRODUCTION

Alternations of Elevation and Climate During Geological Time

Several years ago,[2] I had the honor to give a talk upon "Climate and Evolution" before the Linnæan Society. The subject was then new to me—it was an application to vertebrate paleontology of theories in regard to geological history which had been brought forward by Chamberlin a year or two previously.[3] I have had these concepts more or less in mind ever since, and though I must admit that I am far from having the evidence in shape for final presentation, I desire to submit for general consideration the conclusions thus far reached.

Chamberlin's theories are to-day well known and are year by year gaining a wider acceptance. So far as they pertain to the present subject, they differ from the older prevailing concept of geological climatic conditions

2. Jan. 14, 1902.
3. T. C. Chamberlin: Jour. Geol. 5-8: 1897-1901.

chiefly in that they involve an alternation of climates through the course of geologic time from extremes of warm, moist tropical and uniform, to extremes of cold, arid zonal climates. The former are the results of prolonged base-level erosion and the overflow of large continental areas by shallow seas. The latter are the results of the re-adjustments needed to bring the continents once more into isostatic balance, involving the general lifting of the continents, especially of their borders, the expansion of the continental areas to their utmost limits and the renewal of rapid erosion.

These alternations of conditions are marked by alternations of the prevalent type of formation in the geological series. The uniform base-leveling corresponds to widespread deposits of limestones and in its waning stages with coal formations. The periods of uplift are marked by thick barren formations, often red in color, by indications of arid conditions in salt and gypsum beds and they finally culminate in great extension of glaciers from boreal and high mountain areas.

Chamberlin's text book of geology may be consulted, for the more exact and extended exposition of these theories. The present purpose is to indicate their application to the evolution of land vertebrates.

Permanency of the Ocean Basins

In the first place, we may note that they depend as a fundamental basis on the general permanency of the great ocean basins. The continents have been alternately partly overflowed, separated and insular, or raised to their greatest extent and united largely into a single mass. The great ocean basins have in the main been permanent. This principle is dependent upon the known facts in regard to isostasy. The rocks underlying the oceans are heavier than those underlying the continents, as is proved by the deficiency of gravity measurements in the continents as compared with those in oceanic areas, the deficiency being most marked in certain, mostly high-lying parts of the continents. The conclusion appears unavoidable that in a broad way the present distribution of land and shallow water on the one hand, of deep water on the other, has been substantially unchanged.[4] Changes in past geography have been of two kinds:

(1) The continents have been alternately partly overflowed and then have emerged to the limits of the continental shelf.

(2) Certain lines of unstable conditions have been subject to folding and crumpling, accompanied with great changes of level.

---

4. In this connection, however, the suggestion of Bailey Willis that the present isostatic compensation may be unusually complete must be borne in mind.

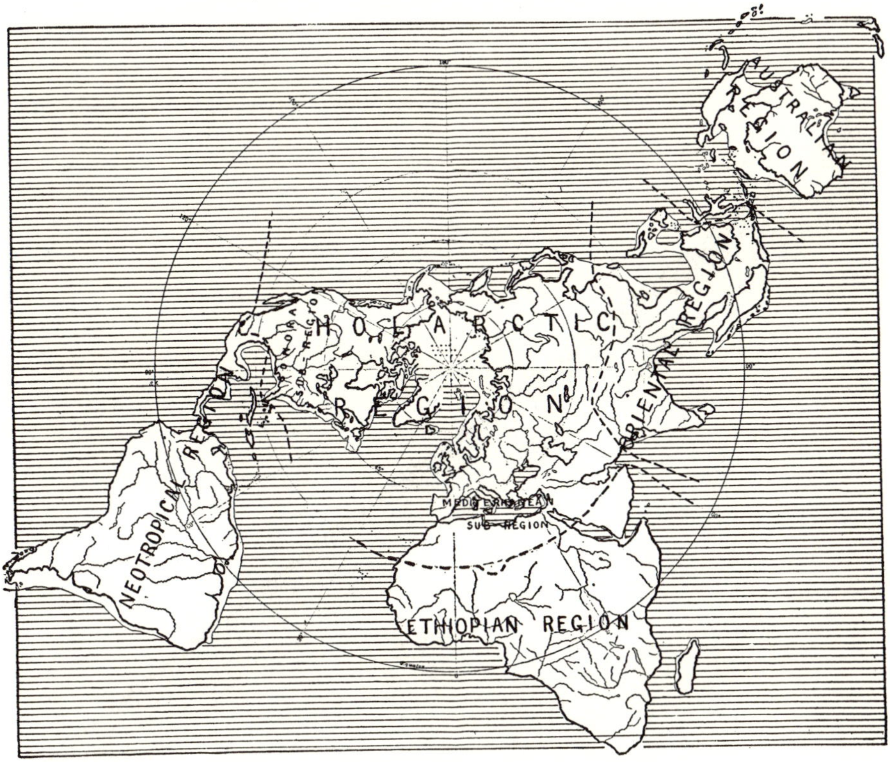

FIGURE 1. Zoölogical regions on north polar projection

The areas within the continental shelf (100-fathom line) are left unshaded. This map represents the true relations of land and water in the northern hemisphere far more correctly than does the usual Mercator projection. The unity of Arctogæa and the direct relation is obvious between the various degrees of isolation of the southern continents and of peculiarity of their faunæ.

Distribution of Land and Water, Present and Past

The present distribution of land and water shows the great land masses located mostly in the northern hemisphere.[5] The land areas, extended to the borders of the continental shelf, form a single great irregular mass with three great projections, South America, Africa and Australasia, radiating out from it into the southern hemisphere. A rise of 600 feet would unite all the land into a single mass.[6] Only New Zealand, Madagascar,

5. It should be observed that the Antarctic continent, according to the latest data available, equals or exceeds any of the other continents in bulk of emerged land; but it is surrounded by deep oceans of vast extent.

6. Australia forms a doubtful exception. The soundings in the Indo-Australasian region are insufficient to determine with certainty whether or not there is any continuous bridge within the 100-fathom line.

the Antilles and numerous small oceanic islands would remain separate. The East Indian islands would be part of the main land. A lowering of 600 feet would isolate North America, South America, Asia, Africa and Australia as separate insular continents. Europe would form a complex of islands and peninsulas much like the East Indies of to-day.

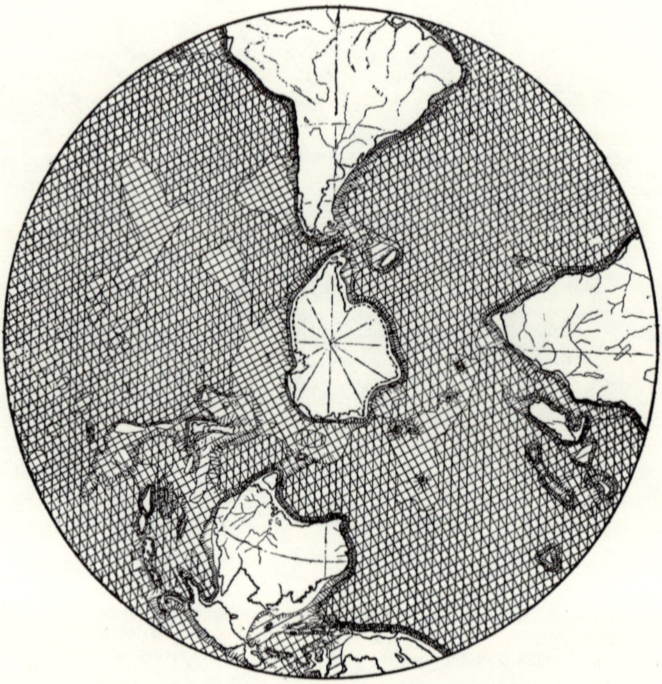

FIGURE 2. The southern continents, south polar projection

Ocean depths of 100–1000, 1000–2000, and over 2000 fathoms indicated by progressive shading. Less than 100 fathoms unshaded. The steep margins of the continental shelf are indicated by hachures. The isolation of the southern continents is in contrast to the unity of the northern land areas.

According to the present theory, we have recently passed through an epoch of maximum continental extension and zonal climate culminating in the Glacial age, marked by great aridity in the equatorial zones, by cold and glaciation towards the poles and in high mountain regions. A much earlier extreme of aridity and glaciation is seen in the Permian,[7]

7. The earlier Paleozoic extremes of aridity—Cambrian and Devonian—do not come within the scope of this discussion.

and less marked extremes at the end of the Trias and at the beginning and end of the Cretaceous. The alternate extremes of warm moist and uniform climates are seen in the early Carboniferous, in the Jurassic, mid-Cretaceous and Eocene. Now the base-leveling and overflow conditions are obviously favorable to the expansion and growth of marine life, especially of the littoral and shallow seas. The conditions of complete emergence of the continents and restriction of the littoral life to the steep and narrow border of the continental shelf will be unfavorable and will tend to what Chamberlin calls restrictive evolution of faunæ. Conversely on land, the great emergence of the continents will tend to expansional evolution and cosmopolitan faunæ, while their partial overflow and isolation will tend to the restriction of land migration and the development of provincial faunæ.

Effects of Alternations of Elevation and Climate upon Evolution of Terrestrial Faunæ

Associated with the isolated continents, we have most tropical uniform conditions of climate, and to this the provincial land faunæ of these periods will be especially adapted. The periods of continental emergence were periods of arid and markedly zonal climate, and the faunæ must adapt themselves to these conditions. Such conditions, while favoring the spread and wide distribution of races, would be unfavorable to abundance of life and the ease with which animals could obtain a living. The animals subjected to them must maintain themselves against the inclemency of nature, the scarcity of food, the variations of temperature, as well as against the competition of rivals and the attacks of enemies. In the moist tropical climatic phase, animals would find food abundant and temperature relatively constant; but the larger percentage of carbonic acid and probably smaller percentage of oxygen in the atmosphere during those phases would tend to sluggishness.

We should expect, therefore, to find in the land life adapted to the arid climatic phase a greater activity and higher development of life, special adaptations to resist violent changes in temperature and specializations fitting them to the open grassy plains and desert life. In the moist tropical phase of land life, we should expect to find adaptations to abundant food, to relatively sluggish life and to the great expanse of swamp and forest vegetation that should characterize such a phase of climate.

The oncoming cold and arid conditions should appear first at the poles and spread towards the temperate and tropical regions. Owing to the distribution of the great land masses, this would involve a general tendency for the great migrations resulting from the emergence of the continents to be outward from the two great northerly masses, and especially from Asia. The tropical and southern continents would be the refuge of the less adaptable and progressive types.

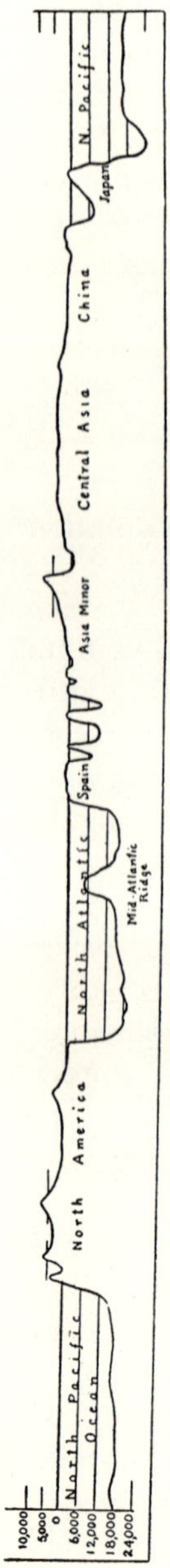

FIGURE 3. Cross-section of continental platforms and ocean basins at 45° north latitude. Vertical scale greatly exaggerated

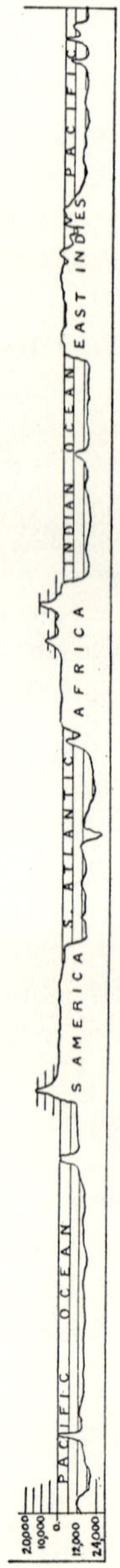

FIGURE 4. Cross-section of continental platforms and ocean basins at the equator. Vertical scale exaggerated about 170 times

This phase of climate should, therefore, favor a higher development and greater activity of land life, while the geographic conditions favor cosmopolitan faunæ. When the climatic pendulum began to reverse its swing, the continents became isolated and their faunæ developed independently; but the dominant animals of these faunæ when first isolated would be those previously developed during the arid phase, and these would readapt themselves to the new conditions of moist and uniform climate, of prevalent forest and swamp and of abundant food.

Comparison with the Paleontological Record

How far do these *a priori* deductions correspond with the facts, as obtained from the geological record? In the first place, we should keep in mind that our record of the land life of the emergence phases is very defective. The sediments of this phase, where deposited along the continental margins, are limited in area, thick and very barren, the conditions of their deposition being generally unfavorable to the preservation of fossils. The sediments of the interior of the continents, river and flood-plain deposits of the Cenozoic era are more widespread and furnish an extensive record of Tertiary and Quaternary land life; but those of the preceding periods of aridity have been re-eroded and carried down to the marginal and littoral areas during the period that has elapsed since they were first deposited. Of the pre-Tertiary epicontinental deposits, only the coast margin, littoral and marine deposits are extensively preserved. That means that the record of Mesozoic and Paleozoic land life as preserved to us is chiefly the record of the coast-swamp and lowland regions and that we know nothing of the life of the upland, except by a rare accidental preservation. In considering the evidences of climatic adaptation during the Mesozoic, this must be kept clearly in mind.

The great mass of evidence in favor of adaptation to progressively arid climate and of dispersal from the northern land regions is derived from the recorded history of the Mammalia during the Tertiary and Quaternary and from comparison of their former and present geographical distribution. It has long been recognized that the present distribution of mammals is due chiefly to migration from the great northern land mass, and the connection of this southward march with progressive refrigeration in the polar regions was made more than a century ago (1778) by Buffon.[8] With a clearer perspective of geologic time and far more exact records, it is clear that most of this deployment and dispersal of the mammalian races has taken place since the Eocene epoch of the Tertiary, although remnants of an older dispersal on the same lines are probably traceable in the present habitat of monotremes, marsupials and primitive insectivores.[((9))]

Interpretation of Supposed Exceptions

There has been a disposition in recent years among students of geographical distribution to lay weight upon certain apparent exceptions to this general rule, where the geological record has not yet afforded evi-

8. See K. v. Zittel, History of Geology and Palæontology, page 43, for a brief summary of Buffon's views on this subject. The theory has been more fully presented by many subsequent writers. In recent years, it has been very ably set forth in its relations to Tertiary Mammalia by Dr. J. L. Wortman (Amer. Jour. Sci. 1903). A very readable little pamphlet by G. Hilton Scribner, entitled "Where Did Life Begin", 1884, while totally deficient in geological perspective, sets forth very clearly the diverse effect upon migration of the general trend of the great mountain system, north and south in the New World, east and west in the Old. Alfred Russell Wallace is, I believe, usually regarded as the foremost exponent of this theory on the distributional side; but it is scarcely necessary to catalogue the principal exponents of a view so long and so generally held.

((9. See Ernst Schwarz, Acta Zoölogica 5: 393-423. 1924. Regards the southern continents as the principal centres of evolution of mammals, and Holarctica as a mere highway of inter-communication. Not to be taken seriously.))

dence to support the northerly origin of certain groups now limited to the southern continents or to the tropics and to infer various equatorial or southern continental connections during or previous to the Tertiary, in order to account for these exceptions.[9a] To these hypotheses, there are several objections:

(1) The evidence for the general permanence of the great ocean basins and their maintenance formerly, as now, by isostatic balance is very strong and direct, and before allowing any exceptions, we should be very sure that no other explanation will serve.

(2) The instances adduced in favor of former equatorial or southern connections are distinctly exceptional cases in the faunæ, which may, in all the cases I have examined, be accounted for by appealing to the imperfection of the geologic record, by parallelism or by the rare accidents of over-sea transportation.

(3) The existence of such land bridges would present the opportunity for migration of other parts or of the whole of certain faunæ, which has evidently not occurred. I can see no good reason why the only animals which availed themselves of such continental bridges should be the ones which might be accounted for in other ways, while those which would furnish conclusive proof are invariably absent.

(4) Many students of geographic distribution proceed on what appear to me to be wholly false premises. They assume that the habitat of the most primitive living member of a race is the original habitat of the race, the most advanced forms inhabiting the limit of its migration. It seems to me that we should assume directly the reverse of this.

Principles of Dispersal

Whatever agencies may be assigned as the cause of evolution of a race, it should be at first most progressive at its point of original dispersal, and it will continue this progress at that point in response to whatever stimulus originally caused it and spread out in successive waves of migration, each wave a stage higher than the previous one. At any one time, therefore, the most advanced stages should be nearest the center of dispersal, the most conservative stages farthest from it. It is not in Australia that we should look for the ancestry of man, but in Asia.

In the same way, in considering the evidence from extinct species as to the center of dispersal of a race, it has frequently been assumed that the region where the most primitive member of a race has been found should be regarded as the source of the race, although in some instances more advanced species of the same race were living at the same time in other regions. The discovery of very primitive sirenians in Egypt while at the same time much more advanced sirenians were living in Europe has been regarded as evidence that Africa was the center of dispersal of this order.

9a. The distinguished Argentine paleontologist, Florentino Ameghino, has for twenty years past advocated a theory the direct opposite to that currently held, and he would derive practically all groups of mammalia from a South American center of dispersal. The evidence for and objections to this theory will be discussed in the sequel.

It is to my mind good evidence that it was not. It is very common to see references to the African facies of the Miocene or Pliocene mammals of Europe; but it is much more correct to say that the modern African fauna is of Tertiary aspect and is in large part the late Tertiary fauna of the northern world, driven southward by climatic change and the competition of higher types.

The chief arguments advanced in support of the method here criticized appear to be that the modification of a race is due to the changes in its environment and that the primitive species are altered more and more as they spread out or migrate into a new environment; but, assuming that a species is the product of its environment, the conclusions drawn would only hold true if the environment remained constant. This is assuredly not the case, and if it were there would be no cause left for the species to change its range. In fact, it is the environment itself, biotic as well as physical, that migrates, and the primitive species are those which have followed it, while those which remained have had to adapt themselves to a new environment and become altered thereby. Probably, it is never the case that the environment of the marginal species is an absolute replica of the older environment of the race. In many cases, it must be profoundly modified by its invasion of new regions, and there are many features in the evolution of a race which appear to be only partly, if at all, dependent on environmental change. But to assume that the present habitat of the most generalized members of a group, or the region where it is now most abundant, is the center from which its migrations took place in former times appears to me wholly illogical and, if applied to the higher animals as it has been to fishes and invertebrates, it would lead to results absolutely at variance with the known facts of the geologic record.

Review of the Evolution of Vertebrate Life

To my mind, this hypothesis of the evolution of land life in adaptation to recurrent periods of aridity supplies a satisfactory background of cause for the whole evolution of the higher vertebrates.

We may set aside earlier periods of aridity and continental extension signalized by the development of invertebrate land types, whose early terrestrial adaptation is wholly hypothetical, since the known portion of their history is so small and so remote from their origin that we cannot project it backwards with any sort of exactness. As Barrell has pointed out, the arid period of the late Devonian coincides with the probable time of the first adaptation of vertebrates to terrestrial life. In the arid period of the Permian, we see the conditions more clearly prevalent which favored a much more extensive development of land life, and this period marks the rise and early differentiation of the Reptilia. That reptiles first differentiated from amphibia as a dry-land adaptation seems to be obvious; that the period of their rise corresponded with the greatest extreme of aridity, continental emergence and glaciation between Cambrian and Quaternary would, I think, be also generally admitted. The dominant order of land reptiles up to the close of the Mesozoic was the dinosaurs, preëminently a dry-land adaptation in their inception, since their most marked characteristic lies in their long limbs, bipedal progression and general parallelism in proportions and structure to the large ground-birds of modern times, which are to-day peculiarly inhabitants of arid regions. The relationship and origin of the more specialized, mostly gigantic, dinosaurs of the later Mesozoic can be best explained by regarding them as a succession of derivatives from smaller and more lightly constructed upland dinosaurs, mostly unknown to us, the larger and more specialized types being re-adapted to a swamp life and inhabiting the coast marshes whose sediments are still preserved, while the more direct line of dinosaurian evolution inhabited the uplands, where the sediments, if such were deposited, have long since been removed by erosion, and the fauna is consequently unknown to us, except by inference. It is quite impossible to trace the evolution of the dinosaurian phyla through the same nearly direct series of known forms as can be done in the phyla of Tertiary mammals. But I may observe that if our knowledge of the Tertiary sediments were limited to the coastal swamp deposits,—if in this country, for instance, we knew only the Tertiary of the Atlantic and Gulf coasts,—we would be equally at a loss for any direct ancestral series illustrating the evolution of the Mammalia.

The same explanation, namely, that the geological record in the Mesozoic is defective where its evidence would be most direct as to the evolution of land vertebrates, applies both to birds and to mammals, but especially to the former. The exceeding scantiness of fossil birds and mammals during the Mesozoic and their apparently sudden appearance in the record, already well deployed, is often explained by supposing them to have evolved mainly in some continent not yet investigated. It appears to me that a simpler and more probable explanation lies in the fact that the formations of the interior of the Mesozoic continents have in general not been preserved and that this facies of the Mesozoic faunæ is consequently unknown to us.

It may be objected that remains of dry-land animals would be brought down by rivers and deposited in their deltas and thus preserved to our day. This may, of course, occur in exceptional cases. How rare is the exception, we may judge from the exceeding rarity of remains of land animals in true marine deposits, where the chances for their preservation should be almost equally great.

In marked contrast with the evolutionary record among dinosaurs, stands the record of development of the non-marine crocodiles and chelonians, whose normal habitat was the swamp regions and whose more direct evolution is in consequence recorded since the Mesozoic. Remaining in a constant environment, they evolved but little, though their abundance and geographical distribution varied.

Throughout all the evolutionary history of the vertebrates, we see numerous examples of races which, having become adapted to a higher plane of life, have re-invaded a lower plane. In each instance, the higher organization and greater activity acquired in the higher plane have caused them to become dominant, increase rapidly in size and spread widely in the absence of efficient competition. Thus we find various groups of marine reptiles appearing with apparent suddenness in the Mesozoic, becoming very abundant and of gigantic size, spreading very widely and then being replaced by new invasions from the land instead of evolving further in their new habitat. The ichthyosaurs, plesiosaurs, mosasaurs, sea-crocodiles, sea-turtles, are examples of this sort among reptiles; the cetaceans and seals among mammals. These invasions from a higher to a lower plane of active life have been very frequent, so that their recognition is necessary in tracing evolutionary series. The converse movement from a lower to a higher plane, as from aquatic to amphibious, from amphibious to terrestrial, from terrestrial to arboreal, or aërial, have been slow, difficult and for the most part have occurred but once or twice in the geological history of vertebrate life. The higher field once occupied, the lower adaptation was handicapped in its attempts to rise.

IMPERFECTION OF THE GEOLOGICAL RECORD

Everyone is familiar with Darwin's classic illustration of the imperfection of the geological record;[10] but I doubt whether the majority of paleontologists realize how very imperfect our record is, even to-day. We know more about fossil mammals in proportion to their modern numbers than about any other of the larger groups of land animals; yet the num-

10. In the Origin of Species, at the end of Chapter X.

ber of species of which we have any adequate knowledge is but a minute fraction of the number which must have lived since the class first came into existence. Were it not so, the fossil species would vastly outnumber the living forms; as it is, they form a small minority. Moreover, the greater number of recorded fossil species are hardly more than *nomina nuda,* each known from a single fragmentary jaw, a tooth, a scale, a broken bone, indicating indeed that an animal otherwise unknown lived at a certain time in a certain locality but giving very little information as to its entire structure, its habits, its geographical and geological range. The relationships of these imperfectly known species, provisionally stated by the describers and adopted without query by subsequent writers, are one of the most fertile sources of error in paleontological theories.

Mammals undoubtedly existed during the entire Mesozoic, an era about three times as long as the Cenozoic. Two thirds of their evolution must have taken place during that time; and by the end of it, the principal modern orders were already defined. But we have not a skeleton, or even a skull of a single Mesozoic mammal.[11] Two jaws and a few teeth from the Triassic, a number of more or less fragmentary jaws[[11a]] from the upper Jurassic and various teeth and fragments of jaws from the uppermost Cretaceous represent the sum total of our real knowledge of the first two thirds of the evolutionary history of the Mammalia. The rest is theory and hypothesis.

Assuredly, we have no right to assume that the few species which have been founded upon these fossil remains represent at all adequately the number and variety of mammals that lived during the Mesozoic; nor can we even suppose that they fairly represent them. Only two[12] of the numerous phyla of early Tertiary mammals can be at all directly derived from known Mesozoic ancestors. The rest are descended from unknown forms. We may suppose, from the evidence at hand, that the known Jurassic and Cretaceous mammals were arboreal swamp-dwellers and that the chief reason why we know so little of the Mesozoic mammals is that the deposits of the upland regions where they chiefly lived have not been conserved to our day, or at all events have not been recognized and sufficiently explored for fossils.

In the Tertiary, mammals suddenly spring into (apparent) prominence,

---

11. Setting aside *Tritylodon* as of doubtful affinities.

[The American Museum Asiatic Expeditions discovered several insectivore skulls in the Cretaceous of Mongolia. These have been described by Gregory and Simpson, and by Simpson.—E.H.C.]

[11a. Simpson has shown that the jaws of *Dromatherium* and *Microconodon* are of reptilian affinities.—E.H.C.]

12. Plagiaulacidæ and Didelphyidæ.

mainly, it may be assumed, because the fluviatile and eolian formations of the Cenozoic still exist in many localities, although they are being rapidly eroded and carried down to the coastal swamp and sea margin areas of deposition. Epicontinental deposits of Eocene age are rare and scattered, and our knowledge of Eocene mammals is obtained from only a few localities and largely from fragmentary specimens. Through the following Tertiary epochs, these deposits become progressively more extensive and abundant, and our knowledge of fossil mammals is correspondingly greater. Finally, in the Quaternary, they form a mantle over most of the earth's surface, and the fossil mammals are so well known and so many specimens from so many localities have been found that we can get a fairly accurate idea as to the range of many species, not merely as discovered in one or another continent, but as to what parts of that continent they inhabited.

If our knowledge of fossil mammals is incomplete, that of fossil birds is very much more fragmentary. They probably came into existence at about the same time as mammals, but the early stages of their evolution are even more obscure, and comparison of the living members of the class affords less evidence than with mammals as to their source and course of progress. They are even rarer than mammals in the Mesozoic. Two skeletons and a feather from the Jurassic of Bavaria, a number of skeletons and fragments from the late Cretaceous of Kansas and a few fragments of the skeleton from Cretaceous formations in New Jersey and Europe,—these are all we know of a class which was probably very large and varied during the Mesozoic. Our knowledge of Tertiary and Quaternary birds is much more extensive, but it bears no comparison to our acquaintance with Tertiary mammals, and the materials on which it is based are for the most part very fragmentary, their identification often questionable. We may say, however, that Mesozoic birds are more completely known than Mesozoic mammals; that is to say, we know the entire skeleton of two or three, and in consequence can estimate their affinities more certainly and exactly. On the contrary, the fragmentary remains of Cenozoic birds make our estimates of their affinities proportionately uncertain and inexact.

The Reptilia are a more ancient class than either birds or mammals and include the ancestral types of both. Our knowledge of fossil reptiles, in comparison with their probable numbers and variety, past and present, is much less than with mammals, more than with birds. We cannot, as with Tertiary mammals, reconstruct approximate evolutionary phyla of the several races from known fossil forms; yet the evidence is sufficient to give a reasonable basis for inferential phyla of some degree of exactitude among many of the Mesozoic and Tertiary reptiles. But the origin of the Reptilia, like that of the Mammalia, is wrapped in obscurity, and the interrelationship of the more ancient groups is a puzzle not yet solved. We have a fairly extensive acquaintance with the Reptilia of certain habitats at certain epochs; but there were evidently long intervening periods and important faunal facies of which we know nothing or next to nothing.

The Amphibia are not a very important group at present and are almost unknown as fossils, except for the so-called armored amphibians or Stegocephalia, whose relations to the modern frogs, toads and salamanders are still far from clear. This ancient group was abundant and varied in Carboniferous, Permian and Triassic times and is supposed to have given rise to the Reptilia; but the relationship has not been satisfactorily demonstrated by fossils, nor is there direct evidence of the interrelationship of the several groups of stegocephalians.

A wide gap separates the oldest four-footed vertebrates from any known fishes, living or extinct.[12a]

ZOÖLOGICAL REGIONS, PAST AND PRESENT

The zoölogical divisions of the land surface of the earth are given by Lydekker[13] as follows:

1. Australian	Region	Notogæic Realm
Polynesian	"	
Hawaiian	"	
2. Austro-Malayan	"	
3. Neotropical	"	Neogæic "
4. Malagasy	"	Arctogæic "
5. Ethiopian	"	
6. Oriental	"	
7. Holarctic	"	
8. Sonoran	"	

The Polynesian and Hawaiian regions have played no material part in the evolution of mammalian faunas and do not call for any special consideration here. The limits of the remaining regions are shown on the accompanying map ((page 5)). The eight principal "regions" are by no means equally distinct, and their combination into three "realms" does

[12a. Säve-Söderbergh's descriptions of Devonian amphibians from Greenland have materially reduced this distinction.—E.H.C.]

13. Richard Lydekker: A Geographical History of Mammals. 1896. This is a modification of the regions proposed by Sclater in 1858 (Jour. Proc. Linn. Soc. 2: 130-146) and adopted by Wallace in 1876 (Geographical Distribution of Animals).

not remove this defect. Of the five included in Arctogæa, the Sonoran is closest, the Malagasy and Ethiopian farthest removed from the central Holarctic region, if we take into account both the recent and extinct faunæ. The true relations of the several regions might perhaps be better represented thus:

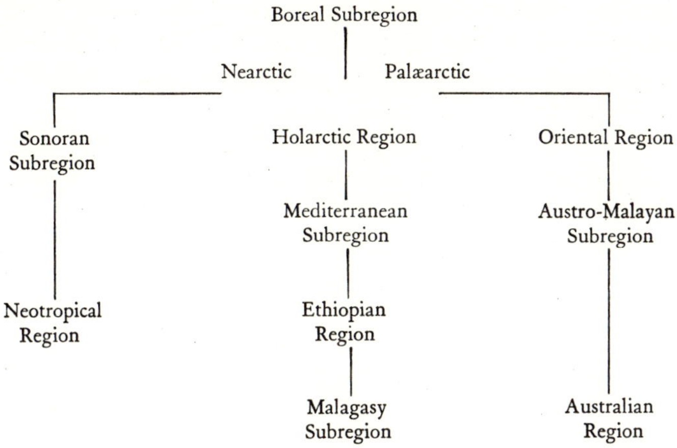

	NEOTROPICAL	HOLARCTIC	ORIENTAL	ETHIOPIAN	AUSTRALIAN
MODERN	Holarctic fauna dominant. A few survivors of the autochthonous fauna	Man cosmopolitan Higher Ruminants Myomorph Rodents Dogs, Cats, Bears.	Antelopes, Elephants Cattle, Deer, Rhinoceros (Fauna of Pliocene and Pleistocene Holarctica)	Antelopes, Horses, Elephants, Rhinoceros	Herbivorous and Carnivorous Marsupials A few Placentals
PLEISTOCENE	Giant Edentates. Macrauchenia and Toxodon. Holarctic Carnivores & Ungulates	Man. Modern genera of Placental Mammals Mastodons, Elephants, Horses, Rhinoceroses.	Elephants, Cattle, Rhinoceroses, Horses, Camels.	Modern? genera of African mammals	Herbivorous and Carnivorous Marsupials.
PLIOCENE	Autochthonous fauna dominant. Earliest invasion of Holarctic fauna.	Mostly modern genera of Placental Mammals. Elephants & Mastodons Rhinoceroses. Monkeys	Mastodons, Elephants, Rhinoceroses, Giraffoid Ruminants 3-Toed Horses etc.		
MIOCENE	Peculiar types of Ungulates. Edentates Marsupial Carnivores	Modern families of Placental Mammals. Mastodons. Monkeys	Primitive Ruminants Mastodons etc	Mastodons? Rhinoceroses	
OLIGOCENE	Evolution of peculiar types of Ungulates etc.	Evolution of modern families of placental mammals		Peculiar types of Ungulates. Invasion of Holarctic fauna	
EOCENE	Primitive Placental Ungulates. Marsupial Carnivores	Modern orders of Placental Mammals (Ungulates, Carnivora, Primates)			
PALEOCENE		Primitive Placental Carnivores and Ungulates			
CRETACIC		Marsupials dominant No Placentals positively known.			

FIGURE 5. Characteristic features of the mammal faunæ in different zoölogical regions at successive epochs of the Cenozoic

The Holarctic region in its broader sense, including the Sonoran and Mediterranean subregions, is bounded by the tropic of Cancer, except where (as in Asia) the dispersal of the fauna from a northern center has been hindered by east and west mountain systems, or (as in America) facilitated by north and south mountain systems. The Sonoran subregion includes most of the United States and northern Mexico; the corresponding subregion in the western half of the Old World is the Mediterranean, including Europe south of the Alps and Pyrenees, part of southwestern Asia and Africa north of the Sahara desert.

The Oriental region corresponds in the eastern part of the Old World to the Mediterranean and Sonoran subregions, but, partly because it includes the great East Indian islands and partly because of the barrier interposed by the Himalayan ranges, it is more clearly differentiated from the Holarctic and may best be regarded as a region of itself.

Austro-Malaya is the debatable ground between the Oriental and the very distinct Australian region; but the consensus of opinion classes it by preference with the Australian. It includes Celebes, the Moluccas, Timor and smaller islands and is separated from the Oriental region by "Wallace's Line."

The Australian region includes Australia, New Guinea and Tasmania and is the most remote and archaic of all the great (continental) regions of the globe. New Zealand is included in the Polynesian (island) region.

The Ethiopian region is connected with the Holarctic by the Mediterranean subregion. It is perhaps more distinct than the Oriental, certainly less so than the Neotropical region. The Malagasy subregion is related not to the modern but to the Tertiary Ethiopian region; its supposed Oriental affinities will be considered later.

The Neotropical region is connected with the typical Holarctic through the Sonoran, as the Ethiopian is through the Mediterranean intermediates; but the relationship is more remote. During the Tertiary, the region was much more distinct than it is now.

In considering the records of past faunæ of one or another of these regions as a guide to the dispersal of different groups, it is very necessary to remember that our records are often chiefly or wholly from a small part of the region, often far from typical.

Our knowledge of Palæarctic faunæ in the early Tertiary is ((chiefly)) from western Europe, an outlying, marginal part, more or less submerged and archipelagic. Its relations to the main body of Palæarctic land life were probably much like those of the East Indian archipelago to the continental portion of the Oriental region. In the later Tertiary and Quaternary, we obtain a broader outlook on the Palæarctic fauna, but even then it is incomplete. ((From the recent discoveries in Mongolia a partial view of the Tertiary faunal succession in central Holarctica is obtained; but it will probably be many years before it is as complete as our records of North American and Western European life.))

In the Oriental region, we know nothing of the land life of the early Tertiary ((except for a glimpse at the Eocene fauna of Burma)), and in the later Tertiary we know only the life of its northern borders, close to the Palæarctic region and doubtless more nearly approximating the Palæarctic fauna then than now, as the Himalayan barrier was less complete.

The result of these two facts will apparently be that the early Tertiary Palæarctic fauna will appear by the record to be less progressive than it really was and that the Tertiary Oriental fauna will appear to be more progressive than it really was. In the Nearctic Tertiary, the record is chiefly confined to the Western plains; we know little of the Canadian Nearctic—presumably more progressive. In the forested regions of the East and South, where we might expect to find primitive survivals, or on the Pacific coast, where we might expect to see stronger Palæarctic influence, our knowledge is very imperfect, although the few available data are in conformity with *a priori* deductions.

In the Neotropical region, our chief dependence is upon the Argentine faunæ which should be both the most progressive and least influenced by Northern immigration.

In the Ethiopian region, we have but a single glimpse of the ((early)) Tertiary land fauna, and that is derived from Egypt, where we might expect to find a transitional fauna, combining true Ethiopian autochthones with immigrants from Palæarctic or northwestern Oriental faunæ. But, since the water barriers to the north of Egypt were more extensive and the desert barrier to the south less developed in the early Tertiary than they are to-day, we should expect that the autochthonic element would be dominant and that Tertiary Egypt belonged to the Ethiopian zoölogical region, although modern Egypt does not.

These may serve as instances of the caution with which the geological record must be used in attempting to estimate the position and source of regional faunæ.

The regions here adopted are based primarily upon the present and past distribution of mammals. Birds, reptiles, amphibians, fresh-water fishes and the various groups of terrestrial invertebrates are not wholly in accord with this arrangement so far as their present distribution is concerned. This is partly because the means and limitations of their dispersal differ, chiefly, as I shall attempt to show, because so little is known of their former distribution.

FORMER BARRIERS AND BRIDGES

The general principle of dispersal on the lines of the present continents is open to an obvious objection. The outlines and connections of the continents were different in former times. The relations of land and water were not the same. In fact, if one depends upon a text-book knowledge of geology he may find authority for an assured belief that they were fundamentally and altogether different in different geologic periods. It is necessary therefore to point out that the stratigraphic no less than the life record is a defective one, and that the really proven changes in the distribution of land and water are limited to those summarized on page 4. The geotectonic hypotheses so ably and brilliantly elaborated by Suess,[14] Haug[15] and other writers, are not facts but theories, and I must confess to a decidedly skeptical attitude towards some of their conclusions. There are too many gaps in the chain of their arguments; too many known facts with which their conclusions appear to be inconsistent.

The permanency of the continental platforms is indicated by the absence of abyssal deposits in their sedimentary succession wherever this has been adequately studied. The platforms have been extensively overflowed by shallow seas, but such submergences were temporary, and intervening periods of uplift are indicated by gaps in the marine succession. Where the geologic records are fragmentary, widely scattered and imperfectly correlated, there often is a tendency to exaggerate the extent and permanency of such overflows, as also to assume extensive unknown continents to account for the existence of clastic sediments which were more probably derived from unsubmerged adjoining portions of the existing continents. We are apt to assume that great displacements of strata involve correspondingly great changes of level. They do not necessarily; more probably, in most instances, the erosion has kept pace more or less closely with the displacement. Even where great changes of level have occurred, they often have been, and more often may have been, of restricted extent and compensated by opposite changes in regions immediately adjoining, and most of them have had but little extensive or permanent effect on the general configurations and relations of the continental platforms.

The relative permanency of the North American continent is very clearly

14. E. Suess: Antlitz der Erde. 1888-1901.
15. E. Haug: Traité de Géologie. 1912.

brought forward in Schuchert's maps.[16] Yet even here, if one may venture a criticism on so ((learned)) and ((brilliant)) a study, there is a certain loss of conservatism where the outlines run into territory where the evidence is inadequate, as in the Antilles and the Arctic seas. The imperfect data available for the South American continent appear to indicate general conditions very similar to those of its northern neighbor; nor does it appear that Africa and Australia were any less permanent land platforms. Northern Eurasia appears to have been similarly permanent, but across Central Europe and extending southeastwardly to the East Indies lies a broad strip of disturbance where great changes have occurred during later geologic time. But the extent and permanency of the great central sea which is so frequently depicted as interposing a broad ocean between the Holarctic and the Ethiopian and Oriental land masses is by no means certain, especially as regards its eastward extension. I cannot find in the recorded facts proof that it afforded any more continuously effective bar to dispersal along the lines of the present continental relations than did the middle Cretaceous overflow in North America or the early Tertiary one in South America.

Perhaps the most widely accepted departure from the permanency of the ocean basins is the supposed Gondwana Land, invented to account for certain similarities in southern Paleozoic floræ, and since used to account for almost all cases of similarity among southern floræ and faunæ which were not demonstrably due to dispersal from the northern continent. This theory has in its original form gone so long uncontested that it is very generally regarded as incontestable. New discoveries have been interpreted in terms of it, the weakness of the original evidence, the possibility that it might be otherwise interpreted, has been forgotten, and like the Nebular Hypothesis, it has become almost impossible to dislodge it from its place in the affections of the average geologist.

If the distribution of animals be interpreted along the lines here advocated, there is no occasion for a Gondwana Land even in the Paleozoic. But it is chiefly as affecting Mesozoic or Cenozoic dispersal that we are here concerned with it. One may summarize the arguments for it by saying that a considerable number of groups of animals and plants which are absent in the northern world, either living or fossil, are found in the southern continents and some of them in certain oceanic islands as well. Most of the groups are unknown or almost unknown as fossils; those which have any considerable fossil record are steadily being eliminated

---

16. Charles Schuchert: Bull. Geol. Soc. Amer. **20**: 427-606. *pl. 46-101.* 1910.

from the list by the progress of discovery, showing that they or their ancestors did formerly inhabit the northern world. The remaining groups agree with those southern faunal groups which have admittedly come from the north, in being of primitive and archaic type and in that their representatives in the different southern regions are but distantly related, the remoteness being in a very direct proportion to the present isolation of the region.

There are a few instances of exclusively southern types closely related (e. g., *Galaxias*); but, although they have been cited in corroboration of the evidence from the groups above mentioned, they are in fact, if thus interpreted, directly contradictory. For the distant relations of the one series is interpreted to mean a very ancient connection, but isolation since; while the other series would indicate a very recent connection and earlier isolation. The explanation here lies not in a northern ancestry, but that the ocean does not form an impassable barrier to their dispersal. This has been proven in the case of *Galaxias;* it is probably the explanation of all similar distributions.

The relations of the *Glossopteris* flora are a different and far more complex problem of distribution. The clue to its interpretation lies perhaps in its association with Permian glaciation; but it is outside the limits of the present essay and will not be discussed here.

REGIONAL CORRELATION

The geological correlation of widely distant formations is so intimately bound up with problems of geographical dispersal and migration that the two series of problems must needs be studied and solved together. We cannot arrive at a correct understanding of the history and causes of the geographical distribution of animals, present and past, without correct correlation of the geological succession in different regions. Nor have we, up to the present time, any reliable methods of exact correlation in widely distant regions except the comparison of fauna and a consideration of their source and the history of their migration and dispersal. Absolute standards, as of world-wide changes in physical or climatic conditions, may serve in the future to give us broad lines of correlation independent of paleontology; but at present their universality is hypothetical, the exact train of physical phenomena which they entail and the indices by which they may be recognized in the stratigraphic succession are imperfectly known. Paleontology is for the present our sole recourse in correlation.

Probably it will always be our chief dependence, at least in exact and detailed comparison.

Synchronism and Homotaxis

The ordinary methods of paleontologic correlation can be applied with accuracy and certainty only over limited areas of the earth's surface. When applied to far-distant regions, we meet first with the difficulty that there is little identity of faunæ, only an equivalence more or less exact. Nor can we be sure that equivalent or even identical forms were contemporaneous in all parts of the earth. They certainly are not so to-day. The modern land fauna of Australia, as Huxley long ago insisted,[17] is in its broad lines a Mesozoic fauna. Examined in detail, it shows indeed the marks of a long period of independent evolution and specialization. Yet the degree and amount of specialization is far less than that which the faunæ of the northern continents have undergone during the Cenozoic. The modern fauna of the East Indies or of Central Africa has a great deal in common with the later Tertiary faunæ of Europe and northern Asia. Central America and tropical South America bear similar relations to North America. While Huxley's dictum that an older fauna in one region may be homotaxial with a later fauna in another does not apply to the extent of involving identity of all or most of the species, yet it very clearly does apply in a broad way to the land faunæ and probably to a less extent to the marine faunæ as well. The rate at which evolution and differentiation progress varies as between the faunæ of different regions. It varies as between the different constituents of a fauna. Neither the partial identity nor the general equivalence of two faunæ is sufficient to prove them synchronous, except under certain conditions to be considered later.

Another method very generally used in correlation of faunæ which contain little or nothing in common consists in an estimate of their relative antiquity as indicated by the proportion of extinct to surviving species or genera. This also involves the assumption that the rate of progress of evolutionary change is constant in all parts of the earth, at least for members of the same group. But if the rate varies in different regions for the fauna as a whole, we have no reason to believe that it would be constant for common or similar groups.

The practical application of this method is very unsatisfactory. In illustration of this, I may instance the widely divergent views entertained by

---

17. T. H. Huxley: Q. I. G. S. **18**: xl-liv. 1862.

different authorities as to the age of the later geological formations of Argentina in comparison with European standards. Able and authoritative discussions of this problem have appeared within the last few years by Ameghino,[18] Roth,[19] Gaudry,[20] Scott,[21] Hatcher,[22] Ortmann,[23] Stanton, von Ihering, Wilckens, Cossmann, Wiman and others, dealing with the vertebrate and invertebrate fossils and stratigraphic relations of the formations. The field work has been extensive, the collections large, the faunæ are large and varied and in large part well known; but the results are widely discordant. The amount of discordance is indicated by the correlation of the four principal terrestrial formations, as given by Ameghino, Roth, Gaudry and Schlosser.

The correlation of widely distant formations is so intimately bound up with problems of geographic distribution and migration that the two series of problems must be studied and solved together. The methods relied upon by Roth and Ameghino are substantially the same as those generally used by northern authors. Why then do they lead to such discordant results? It is because the data on which they rest prove not contemporaneity but homotaxis. Granting that two faunæ in widely remote regions contain the same proportion of extinct species, granting that they represent equivalent stages of evolutionary progress, they are not thereby shown to be contemporaneous, unless they are at the same distance (measured not in miles but in difficulty of advance) from the main center of dispersal of the fauna which they contain. Very obviously, if the principal center of dispersal of Mammalia was in the Holarctic region, the fossil mammals in southern regions invaded by that northern fauna will appear in their homotaxial relations to be more ancient than they really are. The modern fauna of South America, of Africa, of the Oriental regions, will be in the same stage of evolution as the late Tertiary and Quaternary faunæ of Holarctica. Its species will be more nearly related or equivalent to Pliocene and Pleistocene species of Europe and North

---

18. Fl. Ameghino: "L'Age des Formations Sédimentaires de Patagonie," Anal. Soc. Cient. Argent. **50, 54**: 1-231 of separata. 1903. "Formations Sédimentaires du Crétacé Supérieur et du Tertiaire de Patagonie," Anal. Mus. Nac. Buenos Aires **15**: 1-568. 1907.

19. Santiago Roth: "Beitrag zur Gliederung der Sedimentablagerungen in Patagonien und der Pampasregion." Neues Jahrb. Beil. **26**: 92-150. *pl. 11-17.* 1908.

20. A. Gaudry: "Fossiles de Patagonie, etc." Ann. de Paléont. **1**. 1906.

21. W. B. Scott: Mammalia of the Santa Cruz Beds in Rep. Princ. Univ. Exp. Patagonia **5**. 1903. Int. Cong. Zool. Berne C.-R: 241-247. 1905. A History of the Land Mammals of the Western Hemisphere. 1913.

22. J. B. Hatcher: "On the Geology of Southern Patagonia." Amer. Jour. Sci. **4**: 327-354. 1897. "Sedimentary Rocks of Southern Patagonia," *ibid.* **9**: 89-108. *Ibid.* **15**: 483-486. 1903.

23. A. Ortmann: Tertiary Invertebrates in Rep. Princ. Univ. Exp. Patagonia **4**: 45-332. *pl. 11-39.* 1902.

See for further references the bibliography in Ameghino, 1907, supra, 3-18.

America than to their modern fauna. The late Tertiary mammals of the southern continents will approximate in homotaxis the Middle or Early Tertiary mammals of Holarctica; and the Middle Tertiary southern faunæ will approximate the early Tertiary or late Cretaceous faunæ of the north.

Table I

Correlation of the Four Principal Terrestrial Formations

	Ameghino, 1906	Roth, 1908	Gaudry, 1906	Schlosser, 1912
Pleistocene		Pampean	Pampean	Pampean
Pliocene	Pampean			
Miocene			Santa Cruz	Santa Cruz Pyrotherium
Oligocene		Santa Cruz		
Eocene	Santa Cruz	Pyrotherium	Pyrotherium	Notostylops
Paleocene			Notostylops	
Upper Cretaceous	Pyrotherium	Notostylops		
Lower Cretaceous	Notostylops			

On the other hand, if we believe, as does Dr. Ameghino, that the principal theater of evolution of the mammals lay in the temperate regions of South America, and that the mammal population of the North was derived by migration from that center (by way of Africa across a tropical land bridge not now existing), it will be equally obvious that the southern formations will be more ancient than their homotaxis, impartially considered, would lead us to believe. The result will be to assign to the Cretaceous period those southern faunæ which are homotaxial with the early Eocene of the North; to the Eocene those faunæ which are homotaxial with the Middle Tertiary of the North, and so on.

To a certain extent, the intercalation of marine formations may provide a check on this relationship, but it must be remembered that the same theories of dispersal may also apply to marine faunæ, wholly or in part. Homotaxial marine faunæ may be far from contemporaneous. The chief center of dispersal of marine faunæ may be assumed to be either the equatorial oceans and coasts, the northern, or the southern seas, or both north and south equally. Only when the movements of dispersal are in opposite directions on land and in the seas will the marine faunæ furnish an adequate check on the homotaxis of land faunæ; and in that case the true synchronism must be arrived at by balancing conflicting evidence derived from terrestrial and marine faunal comparisons.

It is true that if we eliminate the idea of faunal dispersal altogether and regard each race of animals as evolving and dispersing independently, governed by its own conditions and causes of change, we may in the present imperfect state of our knowledge lay out various and independent centers of dispersal for different races, whose successive appearance in one or another continent will furnish data for a true correlation. There has been a strong tendency in the last half century to work on this theory, but in the present writer's opinion at least, the supposed evidence in favor of this view is due chiefly to the imperfection of the geologic record, and its very wide acceptance to a lack of appreciation of the underlying causes of evolutionary progress and dispersal.

I do not understand how anyone can reconcile the theory that each race of animals evolves and disperses independently and that the common biotic and physical environment is not a controlling factor, with the plain fact that regional faunæ do exist to-day. The conditions that control the dispersal of one race are largely identical or correlated with those that control the dispersal of others, and every change in these conditions will affect not one race only, but a large part or the whole of a fauna, in a manner and to a degree largely identical, causing similar changes in the range of the fauna.

((There are nevertheless various types of animals and plants of wide geographic range, which have overcome the climatic and other limitations which restrain the dispersal of faunas and have become cosmopolitan. These cosmopolitan types, especially when limited in geologic range, as many of them are, are of prime importance in correlation; and if one can check the time of dispersal of one genus by that of another spreading in an opposite direction, they may afford conclusive proof of the true correlation of faunas of widely distant regions.))

Tertiary Correlation in South America

Before setting forth the evidence as to the dispersal of the mammals, it is necessary to attack a problem which has caused much acrid controversy, namely, the age of the later formations of the Argentine Republic. The difference of opinion among authorities has already been indicated, as also the fact that the true correlation is so intimately related to the direction of migration that the two problems must be settled together. In view of the great and well merited reputation of Dr. Ameghino and the immense array of data which he has marshalled in support of his theories of correlation and phylogeny, it is not surprising that they should find a very considerable acceptance, not in South America alone but elsewhere. Few scientists indeed are disposed to accept his derivation of the horse family from early South American ancestors or of the various families of Carnivora from the same source, for in these and other cases the evidence for northern ancestry is almost universally accepted as convincing; but many writers are willing to accept Ameghino's determination of the age of the Argentine formations, although more critical as to his phylogenetic views.

The two, however, must stand or fall together; and it is precisely because the Equidæ, Procyonidæ, etc., if their generally accepted phylogenies be admitted, afford incontrovertible evidence against the validity of Ameghino's correlations of the formations of the Argentine, that he has been compelled to devise different phylogenies for these cases. Few scientists will be willing to believe Ameghino's assertion that *Merychippus* and its successors in the equine phylum have nothing to do with the Anchitheriinæ which they so closely resemble in teeth, in skull, in feet, in all details of the skeleton, but must be derived from the South American Notohippidæ on the strength of a much more distant resemblance in the second upper molar, unsupported by any near resemblance whatsoever in the remaining teeth or in any points of construction of skull or of skeleton. It is not my intention to present here any detailed refutation of Dr. Ameghino's argument, but to point out that if the northern origin of the Equidæ be accepted, the age of the Pampean and related formations must be far later than that he has assigned to them. The first appearance of true equines in South America is in the Pampean. The three best-known genera are *Equus, Hippidion* and *Onohippidion*. The first might be regarded as of Palearctic origin; the second and third have no Old World predecessors, but may be directly derived from the North American *Pliohippus*. They are, however, much larger and more progressive than *Pliohippus,* and in size, reduction of the lateral digits, etc., are equivalent to *Equus*. We can hardly doubt that they came to South America from North America, nor can I see any practical alternative to believing that *Equus* arrived by the same route. Now, the first appearance of

Equus in North America is at the base of the Pleistocene. In Argentina, it first appears in the middle Pampean. The middle Pampean cannot therefore be older and is presumably younger than Lower Pleistocene. *Hippidion* and *Onohippidion* are found (*fide* Roth) in somewhat older levels; but as they are much advanced over anything in our Middle Pliocene (Blanco), it would seem that their first occurrence in the Pampean must be placed at the top of the Pliocene or preferably in the Lower Pleistocene. I conclude that the Pampean formation approximately represents the Pleistocene epoch.

Beneath the Pampean of Ameghino, but included in it by Roth, are fossiliferous beds in which certain Procyonidæ and Ursidæ are found. If we admit the North American source of these Carnivora, they would indicate Pliocene age for the beds containing them. Dr. Ameghino, who regards them as Oligocene and Miocene, is compelled, therefore, to set aside the North American ancestors of the Procyonidæ and to regard them as of South American origin and the Ursidæ as either autochthonous or arriving in South America from the Old World via Africa. As with the Equidæ, the only shadow of plausibility for such phylogenies lies in the incompleteness and careful limitation of the evidence that is adduced in their behalf. *Phlaocyon* of the North American Miocene, which is intermediate between *Cynodictis* and the Procyonidæ in almost every detail of the perfectly preserved dentition, skull and skeleton is merely[24] "un vrai Canide sans relations avec les Procyonidés," while the South American genera are derived through hypothetical ancestors from the carnivorous marsupials of the Santa Cruz. Here again, Dr. Ameghino is compelled, in defense of his theories of correlation, to adopt these impossible phylogenies, because if the Procyonidæ are of North American origin the Argentine formations are demonstrably of later date than those which he assigns to them. *Phlaocyon* is a far more primitive procyonid than any of the South American genera. ((It is in fact not far removed from *Cynodesmus* and *Cephalogale*.)) If these Argentine genera are derived from the Oligocene *Cynodictis* and related genera of Holarctica, *Phlaocyon* being about half way between the two groups, then their age is indicated as

24. FL. AMEGHINO: Ann. Mus. Nac. Buenos Aires 15: 396. 1906. Dr. von Ihering has since attempted to prove what Ameghino merely asserted. His argument rests upon an untenable interpretation of a single feature in the dentition, ignoring all other characters of teeth, skull and skeleton, and, if true, would involve not only that *Bassariscus* has nothing to do with the Procyonidæ (which he asserts), but also that the Procyonidæ have nothing to do with the Carnivora but are of wholly diverse ancestry.

See H. v. IHERING, Systematik, Verbreitung und Geschichte der sudamerikanischen Raübthiere. Archiv f. Naturg., 76 Jahrg. 1: 113-179. 1910.

Pliocene, not as Oligocene or Miocene. Also with the Ursidæ; to admit them as arriving via North America would compel Ameghino to conclude that their first occurrence in South America in these same sub-Pampean beds must be materially later than the evolution of the phylum in the Palæarctic region (Miocene) and that the genus *Arctotherium* of the true Pampean in South America, unknown in North America until the Pleistocene, indicates, like *Equus,* that the Pampean is a Pleistocene formation.

The distribution of *Smilodon* in North and South America is in exact accord with that of *Arctotherium.* The relations of the South American Proboscidea to those of North America correspond to those of the Equidæ. The Camelidæ, Cervidæ, Canidæ, etc., also support the Pleistocene age of the true Pampean. The Edentata, whose migration appears to have been in the reverse direction, will be discussed later.

In the Santa Cruz fauna, we have not the direct evidence that the Pampean faunæ afford for correlation by means of groups of admittedly northern origin. The evidence has been very fully discussed by Hatcher, Ortmann, Scott and others, and so far as it is based upon the relations and age of associated marine formations, I am not competent to criticize it. The criterion used by Ameghino and Roth, of proportions of extinct to living genera, I regard as untrustworthy, partly for the general reasons already given ((page 23)) and partly because of the personal equation that must always affect the number of genera and species described as new, as compared with those referred to known genera and species. Unless the standards of diversity for genera and species were approximately the same, and in this instance they are certainly very wide apart,[25] the comparison of the proportions of extinct to surviving genera and species in Argentine formations with those of Europe or North America would be misleading.

Perhaps the most important correlation is that of the *Notostylops* fauna, Lower Cretaceous according to Ameghino, Upper Cretaceous according to Roth, Paleocene according to Gaudry, Upper Eocene in Schlosser's view. Here there ((was)) an apparently strong point for Cretaceous age in the ((supposed)) presence of dinosaurs in association with the fossil mammals. Dinosaurs disappeared from the Northern world at the end of the Cretaceous.[26] They are entirely unknown in any Tertiary forma-

25. The European fossil rodents are, for the most part, referred in accordance with the old conservative standards of genera and species, while Ameghino is much inclined to hairsplitting in generic and specific distinctions. Scott in his revision is more conservative, but not so as to equalize the standards in question.

26. The latest dinosaur formations of North America are, however, regarded as Paleocene by Knowlton, Lee, Peale, and other authorities.

tion. Nevertheless, the possibility of their survival into the early Tertiary in South America must be considered.[27] ((Recent field studies by Windhausen, Loomis, von Huene, Riggs, and other competent authorities agree in the conclusion that dinosaurs do not occur in association with the *Notostylops* fauna.))[[28]] ((This)) mammalian fauna is in part related to the Paleocene fauna of Europe and North America and for this reason has been regarded as equivalent. But these genera of Northern affinities are associated with a large number of larger and more progressive genera, structurally derivable, according to the canons of evolutionary development universally accepted by paleontologists, from the more primitive types which are common to the *Notostylops* beds and the Paleocene of the North, and leading apparently into the various specialized groups peculiar to the later South American Tertiaries. These more progressive types are unknown to any northern Tertiary fauna; they appear to be derived from the more primitive group whose affinities are so close to the Puerco, Torrejon and Cernaysian mammals; and they point to the conclusion that the *Notostylops* fauna is in reality decidedly later than the Paleocene, the more primitive group of its fauna being little altered survivals,[29] corresponding to the primitive survivals (Condylarthra, etc.) which are found in the Wasatch and Wind River faunæ of North America. Taking the *Notostylops* fauna as a whole, it appears to me to represent an Eocene stage of development, conditioned by an isolation which began ((with)) the Paleocene and hence prevented the incoming of any Perissodactyla, Artiodactyla or Carnivora from North America.[30]

The age of the *Pyrotherium* beds is much less definitely determinable. Dr. Roth, indeed, doubts the existence of this fauna as distinct. If accepted, it would presumably be intermediate between the *Notostylops* and Santa Cruz faunæ and provisionally referable to the Oligocene.

The sequence of the Argentine faunæ will then be

Pampean (s. s.)	= Pleistocene
Monte Hermoso etc.	= Pliocene
Santa Cruz	= Miocene
Pyrotherium	= ? Oligocene
Notostylops	= Eocene.

27. The same arguments apply to the occurrence of a Mesozoic type of crocodile, *Notosuchus*, in the *Notostylops* fauna.

[28. This problem has recently been thoroughly reviewed by Simpson, who agrees that there is no evidence in support of the supposed association of dinosaurs and mammals in South America. See G. G. SIMPSON: Amer. Mus. Novitates **566**. 1932.—E.H.C.]

29. Little altered, that is to say, so far as the parts known to us are concerned; their adaptation, whatever it was, not involving radical changes in dentition from the primary type.

30. Schlosser (in Zittel's Grundzüge d. Pal., Rev. Ed. 1912) regards the *Notostylops* fauna as Upper Eocene. Scott (History of Mammals of West. Hem.) places it as Eocene.

So far as the correlation of the Pampean and Santa Cruz is concerned, their fossils agree wholly in preservation and degree of petrifaction with those preserved in similar Pleistocene and late Miocene formations, respectively, in the western Plains, and the degree of consolidation of the matrix is the same. We have in the West two fossiliferous formations, the Bridger (Eocene) and John Day (Oligocene), which are, like the Santa Cruz, composed of an andesitic volcanic ash, and similar ash strata are found in different levels of our Western Miocene formations. Now, the Santa Cruz matrix and fossils are very much less consolidated or thoroughly petrified than the Bridger and decidedly less so than the John Day, while they agree very well with the volcanic ash beds in the middle and upper Miocene. As there is no reason to suppose that the rock-making processes work at a different rate in different continents, this evidence is entitled to some consideration. On similar grounds, the Pampean fossils would be referred to middle Pleistocene, and the few fossils that I have seen from Monte Hermoso agree best with Pliocene fossil mammals from North America. I should place no weight on this kind of evidence except when, as in the present instance, the climatic conditions and the origin and method of deposition of the formations are substantially similar.

The foregoing digression is somewhat outside the limits of this discussion. It appears, however, to be necessary to show briefly the reasons on which the age assigned to the South American mammalian faunæ are based. It might, indeed, be logically objected that these correlations are based on the northern origin and migration of certain phyla and cannot, therefore, be used in support of the theories here advocated. But the phyla on which the demonstration rests are so universally admitted to have arisen in the north, and the evidence that they did so is so complete and conclusive, that there is no reasonable alternate to accepting them as such. And if so, the correlations of South American faunæ must be approximately as here stated, a conclusion supported by the wholly independent evidence of the degree of consolidation of the formation and of petrifaction of the fossils contained.

CENTERS OF DISPERSAL

Whether the evolution of a race be regarded as conditioned wholly by the external environment or as partly or chiefly dependent upon (unknown) intrinsic factors, it is admitted by everyone that it did not appear and progress simultaneously and *æquo pede* over the whole surface of the earth, or even over the whole area of a great continent. The successive steps in the progress must appear first in some comparatively limited region, and from that region the new forms must spread out, displacing the old and driving them before them into more distant regions. Whatever be the causes of evolution, we must expect them to act with maximum force in some one region; and so long as the evolution is progressing steadily in one direction, we should expect them to continue to act with maximum force in that region. This point then will be the center of dispersal of the race. At any given period, the most advanced and progressive species of the race will be those inhabiting that region; the most primitive and unprogressive species will be those remote from this center. The remoteness is, of course, not a matter of geographic distance but of inaccessibility to invasion, conditioned by the habitat and facilities for migration and dispersal.

If the environmental conditions in the center of dispersal pass the point of maximum advantage for the race-type that is being developed and become unfavorable to its progress, we should find its highest types arranged in a circle around a central region, which was the former point of dispersal, and the more primitive types arranged in concentric external circles. The central region will be unoccupied, or inhabited by specialized but not higher adaptations.

It would appear obvious that the present geographic distribution of a race must be interpreted in some such way as this by anyone who accepts the modern doctrine of evolution. Yet there are many high authorities on geographic distribution who proceed apparently upon a precisely opposite theory. According to these authors, the distribution center of a race is determined by the habitat of its most primitive species, and the highest and most specialized members of the race are most remote from its center of dispersal. This principle may be true enough so far as concerns the first appearance of a given race, *i. e.,* provided the most primitive species are also the oldest geologically; but it appears to me to be the direct reverse of fact as regards the present distribution, or the distribution at any one epoch of the past. The only ground on which it could be defended would be that the progress of the race is due to its migration, and those members which did not migrate did not progress. But this involves the view that its progressiveness up to the time that its geographical environment changed was due to staying at home, and the same progress after its environment changed was due to not staying at home. It seems to me that the prevalence of this view must be due to some fallacious notions about migration, unconsciously retained, involving a concept of it as analogous to travel in the individual. The successful business man, no doubt, may pack up his baggage and take to traveling, leaving home and going elsewhere and profiting much thereby. Nations have done the same thing, likewise to their advantage. But there is very little analogy here to the zoögeographic migration of species—which is a question of expansion or contraction of range, not directly of transference of habitat, although this may be the final result.

It seems obvious that the conditions which brought about the early progressiveness of the race in a particular locality would, so far as they were external, cause the continued progressiveness of those individuals which remained in that region; so far as they were intrinsic, they would affect the main bulk of the race, the center of its range, more than any outlying parts of it. The present writer is very thoroughly convinced that the whole of evolutionary progress may be interpreted as a response to external stimuli; and intends here to point out what he regards as the most important of these stimuli. It is therefore necessary to point out that these postulates regarding centers of dispersal and migration are not dependent upon the theories to be proved—we are not reasoning in a circle.

OCEANIC AND CONTINENTAL ISLANDS

Faunal Differences Between Oceanic and Continental Islands

One of the strongest arguments for the relative permanency of the deep oceans, especially during Cenozoic time, is afforded by the marked and striking contrast between the faunæ of those large islands which are, and those which are not, included within the continental shelf. The continental islands have the fauna of the continents to which they belong, large as well as small, differing only in the absence of types of recent evolution or of unsuitable adaptation and in the survival of primitive types which have disappeared from the mainland. But no question could be raised as to their former union with the mainland, no other possible solution would explain their fauna. We are compelled to assume the former connection of the British Isles with Europe, of Ceylon with India, of Japan with Korea or Siberia, of Sumatra, Java and Borneo with the Malayan mainland, of the Philippines with Borneo, of New Guinea and Tasmania with Australia, of Newfoundland and Cape Breton with Labrador and Nova Scotia. In each and all of these cases, the evidence is overwhelming, and, with the exceptions cited, the faunal identity is complete.

On the other hand, with all those islands which are separated by deep ocean from the mainland, we find that just that evidence is lacking which would afford convincing proof of former union with the mainland. Their faunæ are widely different from those of the adjoining mainland; they lack just those animals which could not possibly have reached there except by land bridges; they point often to long periods of independent evolution and expansion, and the primary elements of the faunæ of every one of them are such as might possibly at least have reached the island without continental union, whether by accidental transportation, by swimming or by other means.

Take for example the mammals of Sumatra, Java and Borneo. We cannot reasonably suppose that the rhinoceroses, tapirs, deer, wild dogs, felids and numerous other large animals common to them and the adjoining continents reached these islands except by land. They are too large for transportation on "rafts" of vegetation such as occasionally drift to sea from the mouths of tropical rivers. They are dry-land animals not given to swimming long distances. And we would not invoke the agency of man to account for a whole fauna. But most important is the fact that all the animals that we might fairly expect to find there in view of a former land connection are really present.

Contrast with this the fauna of Madagascar.[31] There are no ungulate mammals there, except for the bush-pig, possibly introduced by man (in accord with known customs of the Malays) and a pigmy hippopotamus (now extinct) which might have reached the island by swimming, as hippopotami are known to travel considerable distances by sea from one river mouth to another. The great majority of the unguiculate groups of the mainland are also absent. The only representatives are a few very peculiar carnivores of the family Viverridæ, a peculiar group of insectivores (Centetidæ) and a peculiar group of Cricetine rodents, each apparently evolved on the island from a single type introduced long ago, a species of shrew (Crocidura) of more recent introduction and a variety of bats. There are numerous lemurs and no monkeys there; and the lemurs appear to have radiated out from a single group[32] into a number of peculiar types, two of which, now extinct, paralleled the ungulates and the higher apes in several significant features. The fauna of the island does not resemble the present fauna of Africa, nor can it be derived from

31. A. R. Wallace: Island Life. 381-412. 1881. See also Trouessart, Catalogus Mammalium and Suppl. Quinq.; Lydekker, Geog. Hit. Mam. 211-226. 1896. Lydekker's arguments for continental union are mostly invalidated by more recent discoveries.

32. See W. K. Gregory's studies upon the affinities of the Lemuroidea, forthcoming in Amer. Mus. Bulletin.

[W. K. Gregory: Bull. Geol. Soc. Amer. **26**: 419-446. 1915.—E.H.C.]

any one past fauna, known or inferential, of that continent. The attempt to derive it from the present or from any known or inferential past fauna of India involves still greater difficulties. On the contrary, the Malagasy mammals point to a number of colonizations of the island by single species of animals at different times and by several methods. Of these colonizations, the Centetidæ are the earliest, perhaps pre-Tertiary; the lemurs, rodents and viverrines are derivable from one or more middle Tertiary colonizations; and in both cases the "raft" hypothesis may reasonably be invoked.[33] The hippopotami may have arrived by swimming and the bush-pig and the shrew may have been introduced by man, while the bats may readily have arrived by flight. The extinct ground birds are easily derived from flying birds.

Dr. Arldt,[34] in his discussion of the Malagasy fauna, points out its composite character, derived from several successive invasions. This, I think, is clear enough; but it seems equally clear that these were not faunal invasions due to land connection but sporadic colonizations by a few species all at different times. The characters of the mammalian fauna, both negative and positive, practically exclude the theory of land connections during the Tertiary.

The West Indian islands afford another marked instance. In spite of its nearness to Florida, there are no North American mammals in Cuba, except the manatee,—analogous with the hippopotamus in Madagascar. Nor are the other islands richer in fauna. As also in Madagascar, we have a peculiar and very primitive insectivore *Solenodon* (Cuba and Haiti), a number of peculiar extinct ground-sloths, of which *Megalocnus* is the best known, and which although Pleistocene in age are derivable not from the Pliocene or Pleistocene ground-sloths of North or South America but from the Miocene ground-sloths of Patagonia, and evidently differentiated through a long-continued period of isolated evolution, and a couple of chinchillas—the hutias of the larger islands, the (extinct) *Amblyrhiza* in Anguilla. The *Solenodon* may be referred to a more ancient colonization, the ground-sloths probably arrived during the Miocene, the chinchillas more recently; and the direction of the prevalent ocean currents points out the reason why these are of South American derivation. Those who, like Dr. J. W. Spencer,[35] believe in gigantic elevation

33. The moist tropical conditions of early Tertiary times would favor the formation of such rafts, the small size and arboreal habits of the animals concerned would increase the chances of their being caught on such rafts and the uniform climate and consequently more placid seas would increase the distance over which the raft might be transported before it broke up.

34. THEODORE ARLDT: Entwicklung der Kontinente und ihrer Lebewelt. 119-142. 1907.

35. J. W. SPENCER: "Reconstruction of the Antillean Continent," Bull. Geol. Soc. Amer. **6**: 103-140. 1895.

movements connecting the Antilles with the mainland in Pliocene and Pleistocene would account for the absence of the continental fauna by invoking a subsequent subsidence which drowned out everything else. The improbabilities involved in this hypothesis on stratigraphic and faunal grounds have been pointed out by W. H. Dall, R. T. Hill[36] and others.

Cuba, while near in actual distance to the North American continent, has been comparatively inaccessible to sporadic colonization from that source, on account of the direction of the ocean currents; but colonizations from South or (possibly Central) America have reached it. New Zealand is more remote and inaccessible, and, during the whole Mesozoic and Cenozoic eras, we have evidence of but two colonizations by land vertebrates, neither implying any necessary continental connection. The rock-lizard (*Sphenodon*) may, for aught we know to the contrary, be derived from a marine form; all its early Mesozoic relatives were aquatic, some apparently marine. The few other Reptilia may be best accounted for by sporadic colonizations of later date. The moas are probably derivatives from flying birds.

When we come to the smaller oceanic islands, their poverty of fauna is still more conspicuous. If their fauna is due to sporadic colonization, this should be expected, as the chances are reduced directly in proportion to the smaller length of coastline on which an immigrant might land, as well as by their effective distance from the mainland. The colonization of a group of islands one from another may be due to former land connection and subsequent isolation, or to the same method of accidental transport, subject to the same laws of chance.

It is quite possible that in certain instances the small size and unfavorable environment of islands formerly connected with the continent may account for non-survival of the continental fauna. The Falkland Islands are a case in point; but even here, we find the survivors closely allied to the continental fauna and including types which afford the conclusive proof of continental connection which is uniformly lacking in oceanic islands.[37]

The characteristics of continental and oceanic island faunæ have been very fully and ably elucidated by Wallace (Island Life), and it is intended

---

36. W. H. Dall: "Geological Results of the Study of the Tertiary Fauna of Florida," Trans. Wagn. Inst. 3(6). 1903.

R. T. Hill: "Geological History of the Isthmus of Panama and Portions of Costa Rica," Bull. Mus. Comp. Zoöl. 28: 151-285. 1898.

37. Introduction of *Canis antarcticus* by human agency in prehistoric times is, however, a possible explanation of its occurrence. It is the only alternate to a Pleistocene land connection.

here merely to assert that the progressive increase of our knowledge of the past life of the world tends only to emphasize the distinctions in the source of their faunæ which he has so clearly demonstrated and, so far as my acquaintance with the subject goes, to reduce still further the number of continental connections which he regarded as permissible.

To the argument so often advanced that the transportation of a species across a wide stretch of sea and its survival and success in colonizing a new country in this way is an exceedingly improbable accident, it may be answered that, if we multiply the almost infinitesimal chance of this occurrence during the few centuries of scientific record by the almost infinite duration of geological epochs and periods, we obtain a finite and quite probable chance, which it is perfectly fair to invoke, where the evidence against land invasion is so strong. Furthermore, the fact that continents have not in general been peopled in this way one from another is well accounted for by the fact that species already existed there which filled the place in the environment and by their competition prevented the new form from obtaining a foothold, or greatly reduced the chances thereof. In oceanic islands, however, the favorable environment existed without the animal to fill it. Very often, on account of this lack, some other type was evolved to fill its place; birds being widely distributed on account of their powers of flight have in many oceanic islands developed large terrestrial adaptations to take the place of the absent or scanty mammals.

Natural Rafts and the Probabilities of Over-sea Migration thereby

The following series of facts and assumptions may serve to give some idea of the degree of probability that attaches to the hypothesis of oversea transportation to account for the population of oceanic islands.

(1) Natural rafts have been several times reported as seen over a hundred miles off the mouths of the great tropical rivers such as the Ganges, Amazon, Congo and Orinoco.[38] For one such raft observed, a hundred have probably drifted out that far unseen or unrecorded before breaking up.

(2) The time of such observations covers about three centuries (I set aside the period of rare and occasional exploring voyages). The duration of Cenozoic time may be assumed at ((sixty)) million years (((Barrell's)) estimate).

(3) Living mammals have been occasionally observed in such records of natural rafts. Assume the chance of their occurrence (much greater than of their presence being noticed) at one in a hundred.

(4) Three hundred miles drift would readily reach any of the larger oceanic islands

38. Popular Science Monthly **79**: 303-307. Sept., 1911. Gives the recorded observations of the drift of a natural raft of this sort, covering over a thousand miles of travel.

except New Zealand. Assume as one in ten the probability that the raft drifted in such a direction as to reach dry land within three hundred miles.

(5) In case such animals reached the island shores and the environment afforded them a favorable opening, the propagation of the race would require either two individuals of different sex or a gravid female. Assume the probability of any of the passengers surviving the dangers of landing as one in three (by being drawn in at the mouth of some tidal river or protected inlet), of landing at a point where the environment was sufficiently favorable as one in ten, the chances of two individuals of different sexes being together might be assumed as one in ten, the alternate of a gravid female as one in five. The chance of one of the two happening would be $1/10 + 1/5 = 3/10$. The chance of the species obtaining a foothold would then be $3/10 \times 1/3 \times 1/10 =$ one in a hundred.

If then we allow that ten such cases of natural rafts far out at sea have been reported, we may concede that 1000 have probably occurred in three centuries and ((200,000,000)) during the Cenozoic. Of these rafts, only ((2,000,000)) will have had living mammals[39] upon them, of these only ((200,000)) will have reached land, and in only ((200)) of these cases will the species have established a foothold. This is quite sufficient to cover the dozen or two cases of Mammalia on the larger oceanic islands.

Few of these assumptions can be statistically verified. Yet I think that, on the whole, they do not overstate the probabilities in each case. ((The first is, I am confident, a gross understatement.)) They are intended only as a rough index of the degree of probability that attaches to the method, and to show that the populating of the oceanic islands through over-sea transportation, especially upon natural rafts, is not an explanation to be set aside as too unlikely for consideration.

I have considered the case only in relation to small mammals. With reptiles and invertebrates, the probabilities in the case vary widely in different groups, but in almost every instance they would be considerably greater than with mammals. The chance of transportation and survival would be larger and the geologic time limit in many instances much longer. Wind, birds, small floating drift and other methods of accidental transportation may have played a more important part with invertebrates, although they cannot be invoked to account for the distribution of vertebrates. The much larger variety and wider distribution of infra-mammalian life in oceanic islands is thus quite to be expected. And the extent and limits of such distribution are in obviously direct accord with the opportunities for over-sea transportation in different groups.

On the other hand, the transportation of very large animals in this way may fairly be regarded as a physical impossibility, which could not be multiplied into a probability by any duration of time. The only methods of accounting for such animals would be by evolution *in loco* from small

39. Small reptiles and invertebrates would only rarely be observed, if present.

ancestors, by swimming, by introduction through the agency of man and by actual continental union.

The first hypothesis would involve evolution in an isolated and more or less altered environment and would result in wide structural differences from any continental relatives. The second applies with greater probability to large than to small animals, but, except for animals of more or less aquatic habits and within certain limits of distance, it is an apparent physical impossibility. The third may be either intentional or accidental and should be considered in connection with the known custom among Malays and other races, of taming various captured animals and taking them along on sea-voyages. Its application is, of course, limited to distributional anomalies of late Pleistocene or modern origin. The last hypothesis, where it traverses the doctrine of the permanence of ocean basins, appears to me unnecessary, as I have failed to find a single instance of distribution which cannot reasonably be otherwise explained.

Considerations Affecting Probabilities of Over-sea Migration in Special Cases

The probabilities of over-sea transportation to an oceanic island will obviously be much greater if the island is large, and correspondingly reduced if it be of small size. The distance from the mainland will greatly reduce the chances of such rafts making a landing, for two reasons: first, the chances of survival of the animals are reduced proportionately to the length of their journey (or rather, in a varying relation, which for convenience we may consider as a direct proportion); second, most rafts will be carried out from one or more points along the coast, but not from all points equally (that is to say, from the mouths of one or more great rivers, where the conditions are favorable, seldom from any of the small rivers). If we disregard prevalent winds and currents and consider the rafts as drifting out in all directions the probability of their landing on a given island will be directly proportioned to its length opposite the mainland, inversely to the distance. The probabilities of survival of animals, so far as it depends on the raft holding together, will also be inversely as the number of days exposure to the sea, hence as the distance. Comparing Saint Helena, 1100 miles from Africa and 10 miles in diameter, with Madagascar, 200 miles from Africa and 1000 miles in length, we see that the probabilities of effecting a colonization would be $100 \times 3\frac{1}{2} \times 5\frac{1}{2}$, or 3025 times greater in the case of Madagascar. New Zealand, 800 miles long and 1200 miles from the Australian coast, will receive $8/10 \times 1/6$

$\times$ 1/6, or 1/45 as many colonizations as Madagascar, but 80 $\times$ 11/12 $\times$ 11/12 or 67 times as many as Saint Helena.

I believe that it is to their small size rather than to unfavorable conditions for survival that the poverty of fauna, especially of higher vertebrates, in the smaller oceanic islands is due.

The oceanic currents and prevalent winds do, of course, profoundly modify the above generalities in each individual instance. They have prevented the populating of Cuba from North America, while facilitating invasions from South and Central America. The present set of currents reduces the probability of mammals reaching Madagascar from the African mainland, while increasing the chances of Oriental animals reaching it. It reduces materially the opportunities for Australian fauna to reach New Zealand.

We have no adequate data on which to base theories as to the former set of oceanic currents. A worldwide uniformity of climate would probably reduce the north and south movement of the waters; the east and west element of their motions is conditioned by the rotation of the earth, and its velocity would be reduced proportionately to the north and south movements; so that a more uniform climate would bring about a reduction of velocity rather than change in direction. The third principal conditioning element is the conformation of the continents, and doubtless the flooding of great areas and the opening up of broad though shallow passageways between seas now separated would profoundly modify the surface currents in many regions. The opening of a broad passage between North and South America would allow the Caribbean current to pass into the Pacific instead of being deflected northward and eastward along the shores of the Gulf of Mexico to find an outlet between Cuba and Florida. The absence of this initial part of the Gulf Stream would obviously be unfavorable to North or Central American animals reaching western Cuba. The great equatorial current would sweep across from Africa along the northern coast of South America, and uninterruptedly into the Pacific; transportation from Africa to South America or from South or Central America to the Galapagos Islands would thus be facilitated.

DISPERSAL OF MAMMALIA

Mankind

We may with advantage begin our review of the special evidence in support of our theory with the migration history of man. This is the most recent great migration; it has profoundly affected zoögeographic conditions; it is the one where our data are most complete and accurate; we can perceive its causes and conditions most clearly, and we have a great deal of corroborative evidence in history and tradition.

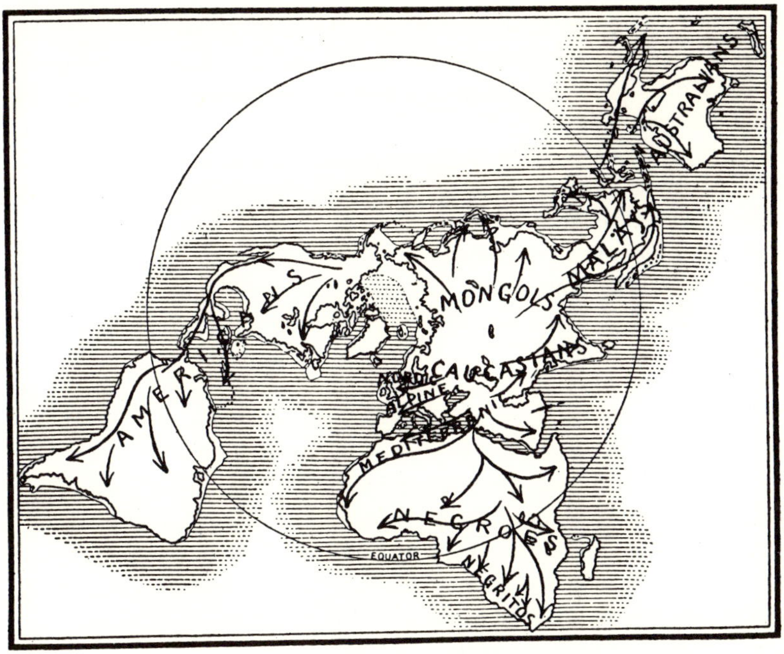

FIGURE 6. Dispersal and distribution of the principal races of man

No attempt is made to indicate anything beyond the broader lines of dispersal.

((Most)) authorities are to-day agreed in placing the center of dispersal of the human race in Asia.[39a] Its more exact location may be differently interpreted, but the consensus of modern opinion would place it probably in or about the great plateau of central Asia. In this region, now barren and sparsely inhabited, are the remains of civilizations perhaps more ancient than any of which we have record. Immediately around its borders lie the regions of the earliest recorded civilizations,—of Chaldea, Asia Minor and Egypt to the westward, of India to the south, of China to the east. From this region came the successive invasions which overflowed Europe in prehistoric, classical and mediæval times, each tribe pressing

[39a. Recent discoveries by Dr. Robert Broom show the presence of very primitive hominids in South Africa.—E.H.C.]

on the borders of those beyond it and in its turn being pressed on from behind. The whole history of India is similar,—of successive invasions pouring down from the north. In the Chinese Empire, the invasions come from the west. In North America, the course of migration was from Alaska, spreading fan-wise to the south and southeast and continuing down along the flanks of the Cordilleras to the farthest extremity of South America. Owing to the facilities for southward migration afforded by the great Cordilleran ranges, the most remote parts of the New World are the forests of Brazil and of northeast South America. In the northern continent, Florida is the most distant from the source of migration.

In Africa, the region north of the Sahara has been overrun by successively higher types from the east. The great desert was a barrier to southward migration, being pierced only by the narrow strip of the Nile valley, from whose head spread out the successive populations of central and southern Africa. The main trend of migration followed the eastern highlands, the valleys of the Niger and Congo being more remote.

In the East Indies, the succession of great islands to the southeast, perhaps more connected formerly than now, formed stepping stones of migration to the distant continent of Australia.

The lowest and most primitive races of men are to be found in Australasia, in the remoter districts of southern India and Ceylon, in the Andaman Islands, in southwest and west central Africa and, as far as the New World is concerned, in northern Brazil. These are the regions most remote, so far as practicable travel-routes are concerned, from Central Asia. A century ago, the present habitat of primitive races was taken to be approximately the primeval home of man. With our present understanding of the conditions and causes of migration, a theory more in accord with tradition and history is generally accepted, and the dispersal center of man is regarded as situated in central or southern Asia. The influence of the old opinion is perhaps seen in the tendency to place this region south of the great Himalayan ridge and in tropical or semi-tropical climate.

This last assumption—that man is primarily adapted to a tropical climate—is, I think, only partly true at best. Its general acceptance is perhaps due, among other reasons, to the supposed relation between loss of hair on the body and the wearing of clothes, the first being regarded as an earlier specialization in an environment of tropical forests, the second as a secondary adaptation resulting from migration to a cold climate. But here, it seems to me, we are putting the cart before the horse. We may more reasonably regard the loss of hair in the human species as a result of wearing clothes and conditioned by this habit, rather than attribute it to any climatic conditions. This view is supported by several points in which the loss of hair in man is differentiated from the partial or complete loss of hair common in tropical animals, the following two being most clearly significant.

(1) It is accompanied by an exceptional and progressive delicacy of skin, quite unsuited to travel in tropical forests. I do not know of any thin-haired or hairless tropical animal whose skin is not more or less thickened for protection against chafing, the attacks of insects, etc.

(2) The loss is most complete on the back and abdomen. The arms and the legs and, in the male, the chest, retain hair much more persistently. This is just what would naturally happen if the loss of hair were due to the wearing of clothes,—at first and for a long time, a skin thrown over the shoulders and tied around the waist. But if the loss of hair were conditioned by climate it should, as it invariably does among animals, disappear first on the under side of the body and the limbs and be retained longest on the back and shoulders.

It will not be questioned that the higher races of man are adapted to a cool-temperate climate, and to an environment rather of open grassy plains than of dense moist forests. In such conditions they reach their highest physical, mental and social attainments. In the tropical and especially in the moist tropical environment, the physique is poor, the death rate is high, it is difficult to work vigorously or continuously, and especial and unusual precautions are necessary for protection from diseases and enemies against which no natural immunity exists and which are absent from the colder and drier environment.

This lack of adaptation to tropical climate is also true, although to a less degree, of the lower races of man. Although from prolonged residence in tropical climate they have acquired a partial immunity from the environment so unfavorable to the newcomer, yet it is by no means complete. The most thoroughly acclimatized race—the negro—reaches his highest physical development not in the great equatorial forests but in the drier and cooler highlands of eastern Africa; and when transported to the temperate United States, the West Coast negro yet finds the environment a more favorable one than that to which his ancestors have been endeavoring for thousands of years to accustom themselves. In tropical South America, the Indians, as Bates long ago remarked, seem very imperfectly acclimatized and suffer severely from the hot moist weather; much more than the negroes, whose adaptation to tropical climate has been a much longer one.

In view of the data obtainable from historical record, from tradition, from the present geographical distribution of higher and lower races of men, from the physical and physiological adaptation of all and especially of the higher races, it seems fair to conclude that the center of dispersal of mankind in prehistoric times was central Asia north of the great Himalayan ranges, and that when by progressive aridity that region became desert it was transferred to the regions bordering it to the east, south and west. We may further assume that the environment in which man primarily evolved was not a moist or tropical climate, but a temperate and more or less arid one, progressively cold and dry during the course of his evolution. In this region and under these conditions, the race first attained a dominance which enabled it to spread out in successive waves of migration to the most remote parts of the earth. The great mountain ranges to the south impeded migration in this direction, while to east and northeast, west and northwest, migration was easy and rapid. Reaching the New World by way of the Alaskan bridge, the long uninterrupted chain of the Cordilleras facilitated migration along their flanks to the farthest limits of South America.

There is little evidence if any, in the New World, of any migrations of inferior races long preceding those of the Amerind tribes, which would seem to have branched off at a moderately high stage in the evolution of mankind. *Per contra,* we find in South Africa, in Australia, in peninsular India and elsewhere, remnants of what must have been an early cycle of migrations. Each group of this early cycle, derived primarily from a different part of the central region of dispersal, has specialized further in proportion to its isolation and yet retains a predominance of the common primitive characters representing the stage of development attained when it left the dispersal center. The populating of Africa by the negroes may be regarded as the latest phase of this early cycle of dispersal, or should perhaps be considered independently.

The later development of the race is conditioned by its splitting in the region of dispersal into an eastern or Mongolian and a western or Caucasian stock. This split was presumably conditioned by the east-west elongation of the dispersal center caused by the facility of expansion in these directions and the mountain barriers to the south. All the eastward migrations from this time on bear a distinctly Mongolian stamp. An early phase of this stage is represented by the population of the New World and the variously mixed Malayan peoples. A later phase appears in the more typical Mongolian races. All the westward migrations, on the other hand, are of Caucasian affinities, this stock splitting, as the region of favorable environment widened out westward, into northeastern or Nordic, southwestern or Mediterranean groups. The peoples of northern Europe are derived from the successive migration waves of the first, those of southern Europe and northeast Africa from the second; the intermediate Alpine stock of central Europe is considered to represent a somewhat older migration allied to the Slavic peoples, who are to-day the principal population of eastern Europe, the latest cycle of Caucasian dispersal.

I have gone into this brief recital of the migration and dispersal history of mankind, not to present anything novel or authoritative, but because we have more evidence, direct and indirect, and more insight into the conditions and causes which controlled its course, than with any other race of mammals. I believe that these controlling causes have been substantially the same in the lower animals as in man and their methods and routes of dispersal largely identical.[40]

Primates

We have seen that the dispersal center of man is in central Asia; that, in the present distribution, the survivors of the earliest cycle are found in Africa, peninsular India, the East Indies and Australia; that the populating of the New World belongs to a later cycle of distribution, and we have no good evidence that the earlier cycle ever reached it; that the dominant migration in the Old World has been east and west, progress to the south being hindered by the transverse mountain system to the south of which more primitive types long survived, while in the New World the dominant line of migration has been to the southward from Alaska, and eastward migration has been slower.

In the living Primates we have survivors of pre-human stages in the evolution of man, specialized to a varying extent in different directions from him, so that they have not come into direct rivalry with him, and have hence survived.

The latest infra-human cycle is represented by the anthropoid apes, surviving to-day in the forests of West Africa and of the East Indies. We may suppose that these are remnants of a cycle of dispersal from a central Asiatic source, but we have no sufficient data to define its extent or time, except as late Tertiary and ((chiefly)) limited to Arctogæa. Nearest to man in intelligence and habits, this cycle has been swept out of existence, except for the few members which were or became adapted, as our own ancestors were not, to tropical forest life. The arboreal habitat may be interpreted

40. One notes, too, the same fallacy in interpreting the data; some authors are disposed to place the center of dispersal of European races or languages in western Europe or in northern Africa because they find there the most primitive surviving races or languages.

as a partial reversion. The doubtful and fragmentary remains of anthropoid apes in the ((late Miocene and)) Pliocene of Europe and northern India are about all that the geological record has to state in regard to the former distribution of this cycle. ((Gregory's researches and conclusions would appear to fix this common ancestral stock as existing in the Miocene, and progressively differentiated during that epoch.[40a]))

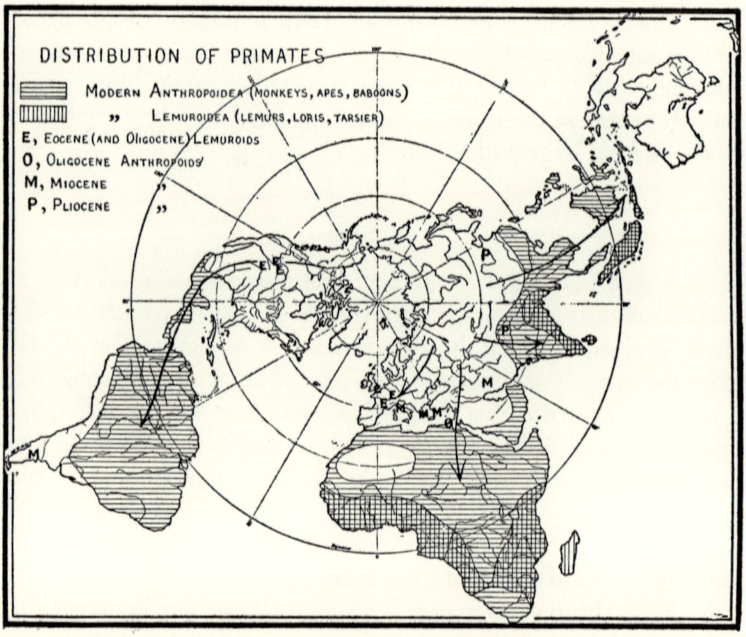

FIGURE 7. Dispersal of the Primates

The marginal position of the modern lemurs, the progressive disappearance of the order from the more central regions which it formerly inhabited are clearly shown.

((The discovery of an upper molar of an anthropoid primate, *Hesperopithecus,* in the Upper Snake Creek bed, associated with a large and well correlated mammal fauna, indicates that this cycle of dispersal did reach the New World although presumably not in sufficient vigor to maintain itself there.[40b]))

The next lower cycle is that of the monkeys and baboons of the Old

((40a. W. K. GREGORY: Bull. Amer. Mus. Nat. Hist. 25: 239-355. 1916.))

[40b. The *Hesperopithecus* tooth, originally considered by Osborn, by Matthew, and by Gregory as belonging to an anthropoid primate, was subsequently shown by Gregory to be a worn cheek tooth of a Pliocene peccary, *Prosthennops.*—E.H.C.]

World, and as a very doubtful early phase, the New World monkeys. The Old World monkeys inhabit most of Africa, India and the East Indies. To the northeast they extend to southern Japan. Closely related forms are found in the late Miocene of central and southern Europe, in the Pliocene of India, in the Pliocene ((and Pleistocene)) of China. These may all be referred to a central Asiatic ((or Oriental-Mediterranean)) source. The dispersal of this cycle must date back at least to the beginning of the Oligocene, at the date of the Fayûm fauna ((of Egypt)) as shown by

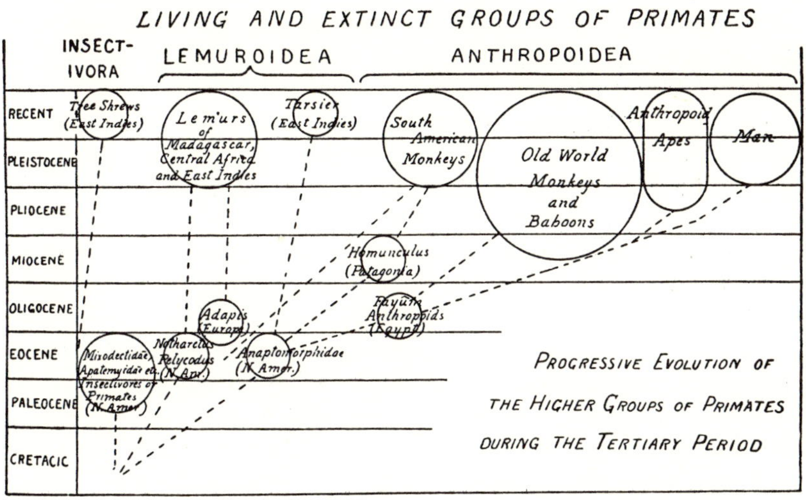

FIGURE 8. Phylogenetic relations of the living and extinct groups of Primates

The circles indicate the size and known geological range of the several groups, the dotted lines their most probable derivation. Their supposed relations to certain Insectivora and intermediate extinct groups are also indicated.

Schlosser's recent discoveries.[41] With the New World monkeys, the evidence seems rather to point to independent evolution in South America from early Tertiary Primates of an Eocene cycle of dispersal. For no remains of Primates have been discovered in any Oligocene or later formation in the United States, while the later Tertiary formations of the Argentine have yielded remains of a number of Primates apparently intermediate between Eocene lemurs and South American monkeys.

The oldest cycle of primate dispersal is that represented by the lemurs. These are now most abundant in Madagascar; a few exist in west and

41. MAX SCHLOSSER: "Beiträge zur Kenntniss der Oligozänen Landsäugethieren aus dem Fayum Aegypten," Beit. zur Pal. u. Geol. Oest-Ung. **24**: 52. 1912.

central Africa, peninsular India and the East Indies. Lemuroid primates, lacking certain specialized characters of modern lemurs but otherwise closely related, and equivalent in stage of development, are found abundantly in the Eocene of Europe and the United States. They are very doubtfully represented in the early Tertiary formations of the Argentine. We know too little of the Tertiary of other parts of the world to make any inference as to the extent of their distribution at that time, or the course of its subsequent changes. They disappear in Europe and North America at the end of the Eocene; in South America, they may have evolved into New World monkeys, while in the Old World they must have given rise to the higher primates. It is reasonably certain that the theater of their evolution was not Europe, and although they are not known in the Oligocene Fayûm fauna of Egypt, we may doubtfully suppose that they had reached that continent at some time during the Eocene. Madagascar most probably received its lemurs from Africa, but it is reasonable to suppose that only a single type, allied to the Eocene Adapidæ, reached the island, and in the favorable environment radiated out into a number of diverse adaptations taking the place of various mammal groups not present in the island fauna.

From the fact that the European and North American lemurs are in an equivalent stage of development, although not very closely related, we may fairly infer that they were derived very early in the Tertiary from an intermediate center of dispersal, presumably Asia north of the Himalayas.

Carnivora

The modern land Carnivora are spread over all the great continents except Australia, where a single species of wild dog, probably introduced by man, is their only representative. They are found equally in all the continental islands (*i. e.,* those included within the continental shelf border), and a few have reached Madagascar and other large oceanic islands.

The order is unquestionably of Holarctic origin. Primitive Carnivora (Creodonta) are abundant in all the earlier Tertiary formations of Europe and North America, one group (Miacidæ) ancestral to the higher Carnivora (Fissipedia), others which became extinct during the Oligocene. True fissipede Carnivora first appear in the Upper Eocene in Europe and North America and differentiate into the diverse modern types through the remainder of the Tertiary. They did not reach South America until the Pliocene, their place being supplied up to that time by carnivorous marsupials. In Australia, their place is still taken by carnivorous marsupials. In Africa, primitive Carnivora (Creodonta) appear in the Oligocene, represented only by the extinct family of hyænodonts, all of them derivable from Eocene hyænodonts of the Holarctic region; but the contemporary Holarctic Fissipedia had not yet reached Africa.

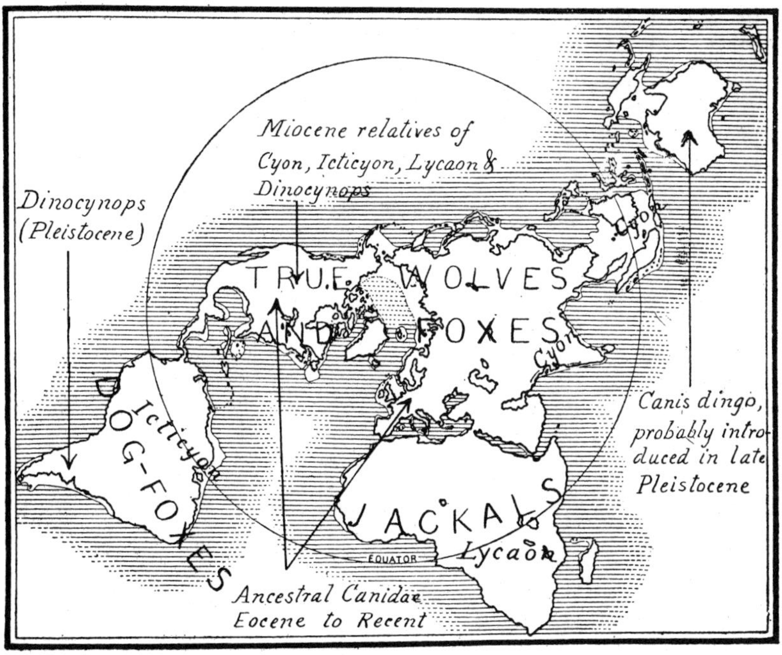

FIGURE 9. Distribution of the modern Canidæ

The jackals (Ethiopian and Oriental) are slightly more primitive than the true wolves and foxes; the Neotropical "Dog-foxes" more distinctly so. *Cyon, Icticyon* and *Lycaon* appear to be dispersed remnants of an aberrant group formerly Holarctic; the ancestry of the more typical Canidæ is also found in Holarctica.

The modern land Carnivora are divided into seven families, each representing one or more broad phyla. The various divergent adaptations of the phyla and secondary re-adaptations of subphyla have brought about an amount of convergence and parallelism which makes it difficult to disentangle or to state accurately the true genetic relationship in any terms of classification. Some of the phyla are Holarctic, others Palæarctic or Nearctic. In all of them, we find the most primitive modern survivors in the tropical regions, the most advanced types in the Holarctic.

CANIDÆ.—The Canidæ are the most cosmopolitan family of the order. It is also the most progressive family in its adaptation to the open plains and arid climate of the great modern continents. The gradual adaptation of the race to these conditions from primitive arboreal forest-living ancestors can be traced through successive stages in the Tertiary formations of Europe and North America, but most completely in the latter country.

TABLE 2

DISTRIBUTION OF THE CANIDÆ

	Neotropical	Nearctic	Palæarctic	Ethiopian	Oriental	Australian
Recent	*Canis* *Icticyon*	*Canis*	*Canis*	*Canis* *Lycaon*	*Canis* *Cyon*	*Canis*
Pleistocene	*Canis* *Dinocynops*	*Canis*	*Canis* *Cyon*			*Canis*
Upper Pliocene						
Lower Pliocene		*Hyaenognathus*				
Upper Miocene		*Aelurodon* *Tephrocyon* *Amphicyon*	*Simocyon*			
Lower Miocene		*Cynodesmus* *Daphœnodon*	*Galecynus* *Amphicyon*			
Upper Oligocene		*Enhydrocyon* *Nothocyon* *Temnocyon*		None[42]		
Lower Oligocene		*Cynodictis* *Daphœnus*				
Upper Eocene		*Procynodictis*	*Cynodictis*		Record inadequate	
Lower Eocene		*Cynoid* *Miacidæ*				

The lengthening of the limbs and their adaptation for swift running, the reduction of the long balancing tail to a short comparatively unimportant organ, the perfection of the shearing and crushing teeth and, especially, the steady increase of brain capacity are the chief lines of progress. While most of the surviving Canidæ conform pretty closely to a single type, we

((42. Fayûm fauna of Egypt, transitional between Palæarctic and Ethiopian. Absence of Canidæ in this fauna indicates *a fortiori* its absence in Oligocene Ethiopia.))

find a tendency among their Tertiary ancestors to branch off on the one hand into more predaceous, on the other into more omnivorous types. Most of these have disappeared, but in the Oriental Ethiopian and Neotropical regions we find in the genera *Cyon, Icticyon* and *Lycaon* survivors of a more predaceous group which is known from the Oligocene and Miocene of the Holarctic region. This group has disappeared from

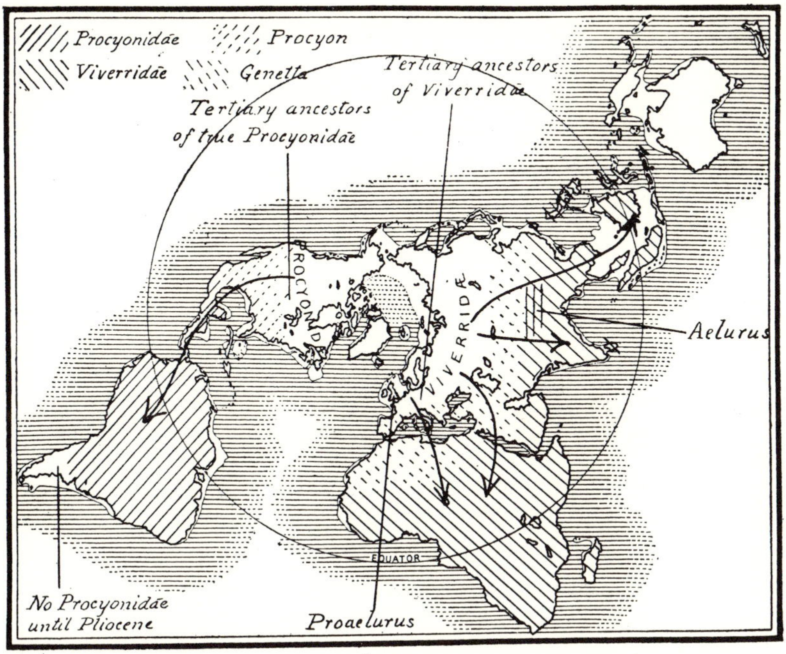

FIGURE 10. Distribution of the Procyonidæ and Viverridæ, formerly Nearctic and Palæarctic, but now surviving chiefly in the peripheral regions

The geographical position of *Aelurus* is anomalous for a member of the Procyonidæ, to which family it is usually referred. Its true affinities, however, are doubtful.

Holarctica by the end of the Tertiary; two or three representatives are found in the Pleistocene of South America. Among the more typical modern dogs, the wolves and foxes are the most progressive types, the jackals slightly less so, the African fennec retains most nearly the primitive long tail, the South African *Otocyon,* while anomalous in possessing an extra molar tooth, is likewise normally primitive in several characters and the Neotropical "dog-foxes" show a marked resemblance in many details to the late Tertiary Canidæ of North America. The fact that the

Canidæ are preëminently adapted to open country and more or less arid climate is of primary importance in explaining their present dominance and cosmopolitanism, their close association with man, their absence from Madagascar and other oceanic islands; and it makes it most probable that the introduction of the dingo to Australia was through human agency although undoubtedly as early as the late Pleistocene. In their adaptation and distribution this family of Carnivora largely parallels the Equidæ among Perissodactyla.

Procyonidæ.—The family Procyonidæ includes a number of omnivo-

Table 3

Distribution of the Procyonidæ

	Neotropical	Holarctic	
		Sonoran	Palearctic
Recent	*Procyon* *Nasua* *Cercoleptes* *Bassaricyon* *Bassariscus*	*Procyon*	*Ælurus* (Affinities questionable)
Pleistocene	(Not recorded)	*Procyon*	((*Ælurus*)) ((*Ælureidodon*))
Pliocene	*Amphinasua* *Pachynasua* *Cyonasua*		*Parailurus* (Affinities questionable)
Miocene	None	((*Probassariscus*)) *Phlaocyon*	
Oligocene	None	*Cynodictis* (Probably ancestral in part)	
Eocene	None	Miacidæ	

rous specializations from the central phylum now represented by the Canidæ. All of them are arboreal, partly retaining and partly reverting to the primitive mode of life in this respect. They are mainly Neotropical, but the raccoon, the most dog-like of the family, survives as far north as the Sonoron region. The panda of the Himalayas is usually placed with Procyonidæ, but its affinity ((to the American genera)) is not very ((close)).[42a]

[42a. See Matthew's annotation on page 54, in which his latest opinion as to the relationships of the panda is set forth. See also the paper by Gregory, cited on the above-listed page, in which the procyonid affinities of the panda are confirmed.—E.H.C.]

Mustelidæ.—Primarily the Mustelidæ represent a more predaceous adaptation than the Canidæ. Their development through the Tertiary in the Holarctic region can be traced almost as completely as that of the dogs. Like the Canidæ (though not as early), they perfected during the later Tertiary a differentiation of the back teeth into shearing and crushing types, and they are equally progressive in brain development but much less so in running powers, retaining to a great extent their primitive forest-living habitat. They are to-day chiefly Holarctic, the most progressive typical mustelids being the martens, weasels, ferrets and wolverenes. Early in the Tertiary there appear divergent side branches, specialized descendants of which survive to-day in the badgers, skunks and otters of the northern world, the intermediate forms being now extinct or confined to India and Africa.

Ursidæ.—The bears are regarded by many paleontologists as an offshoot from the Canidæ, but, on structural evidence, they appear to be related rather to the Mustelidæ. Their distribution indicates derivation from a Palæarctic source.

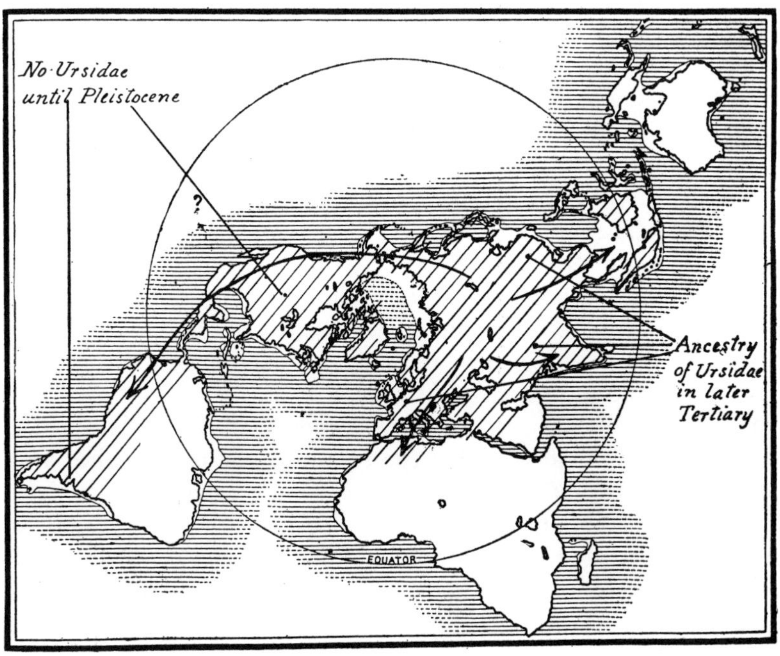

Figure 11. Distribution of the Ursidæ, Pleistocene and Recent

The group appears to have dispersed from a Palæarctic center, its Tertiary ancestral series being found in Europe and in the Pliocene of India and China.

((The most primitive bears (*Ursavus, Hemicyon*) appear first in the Miocene of Europe and America (Sansan, Paunee Creek, Santa Fé marls). In the Pliocene *Hyænarctos* succeeds *Hemicyon* and true *Ursus* replaces *Ursavus*. In the Pleistocene, *Ursus* is Holarctic, while the somewhat more primitive *Arctotherium* is Nearctic and Neotropical, and the modern fauna shows *Ursus*, Holarctic; *Thalarctos*, a boreal specialization; the Neotropical *Tremarctos*, more primitive and closely related to *Arctotherium*, and the Oriental sun bear and sloth bear are partly primitive, partly aberrant. The Thibetan *Aeluropus* is perhaps a survivor of the *Hemicyon-Hyænarctos* branch of the family. Its relationship to the typical bears is not close.))[(42b)]

VIVERRIDÆ.—The Viverridæ are now almost exclusively Oriental and Ethiopian and have conserved the primitive type more than any other Carnivora, except some of the Procyonidæ which have a somewhat corresponding geographic position in the New World. The three most progressive genera, *Herpestes, Genetta* and *Viverra*, survive to-day along the southern borders of the Palæarctic region; the remainder are Ethiopian or Oriental, the most primitive living genera being west African and East Indian. *Herpestes* and *Viverra* occur in the Oligocene and Miocene of Germany and France, and more primitive extinct genera in the Upper Eocene of Europe.

The primitive character of the viverrines is especially seen in their imperfect differentiation of shearing and crushing back teeth, their rather short limbs, long bodies, long tails and relatively small brain capacity.

HYÆNIDÆ.—The family Hyænidæ is generally regarded as a specialized offshoot from the Viverridæ and is apparently connected with the European Miocene viverrids by a series of intermediate forms. The latest development of the race, the genus *Hyæna*, inhabited Europe and Central Asia and China in the Pliocene and Pleistocene but is now found only in India, Africa and southwestern Asia.

FELIDÆ.—The Felidæ are almost as cosmopolitan as the dogs and are even more uniform in type, the cheetah being the only marked living variant. A notably different specialization is shown in the extinct machærodonts or sabre-tooth tigers, and in the Tertiary sequence in Europe and America we find approximate genetic series, parallel in the two countries, by which the true cats and machærodonts converge towards a common primitive type, in which the upper canines are moderately elongated.

((42b. *Vide* E. Ray Lankester: Trans. Linn. Soc. London II. **8** (**6**): 163-172. 1901.)) [See also a recent study on the phylogenetic relationships of the giant panda by W. K. Gregory: Amer. Mus. Novitates **878**. 1936.—E.H.C.]

According to this phylogeny, the clouded tiger of Sumatra and Java is the most primitive living felid, while the double series in Europe on one hand and North America on the other, would indicate northern Asia as the center of dispersal of the race. The range of some of the modern species is very great. The puma extends in the New World from Alaska to Patagonia, the tiger in the Old World from Manchuria to Java. We may note, however, that the tiger is regarded by Blanford as a recent immigrant into southern India; while, on the other hand, it is known that the northern range of the lion has been progressively restricted during prehistoric and historic times from northern Europe to its present limits of southwestern Asia and Africa.

Creodonta

((The creodonts are the ancestral group from which the later Carnivora arose. They include, besides the Miacidæ, ancestral to the fissipede and probably to the pinnipede Carnivora, a number of aberrant extinct families, which partly paralleled the later diverse specializations of true Carnivora. Among these are the Arctocyonidæ, paralleling the bears, raccoons, the Oxyænidæ, paralleling the Mustelidæ, and the Hyænodontidæ, combining characteristics of dogs, cats, and hyænas, but more nearly parallel to the carnivorous marsupials than to any other group. So far as known, they are all Holarctic in the Eocene, and have all become extinct at the end of this epoch save for the Miacidæ, specialized into the several existing families of Carnivora, and the Hyænodontidæ which survive in Holarctica until the Middle Oligocene, in Ethiopia until the Lower Miocene or later—*Metapterodon* Strömer 1923 from S. W. Africa, and in India until the Upper Miocene—*Dissopsalis* Pilgrim, from the Chinji zone of the Siwalik beds.))

Pinnipedia

When dealing with littoral and marine mammals we must expect to find the conditions of their evolution somewhat different. If the hypothesis be valid that the progressive refrigeration of the polar regions was the dominant cause of evolutionary progress and geographic dispersal, an examination of the map will show that the Arctic-North Atlantic basin affords the most favorable region. The Arctic basin centers around the pole, and a broad shelf of shallow water encircles it, extending as far south as latitude 45°. The North Pacific basin was closed to the northward by the Alaskan land bridge during a large part if not all of the

Tertiary, and its shores plunge suddenly to great depths, margined by high mountains, affording less opportunity for expansional evolution of the littoral fauna. The Antarctic continent appears equally unfavorable, and dispersal from that center would also be hindered by the broad stretches of ocean.

We may expect, therefore, to find the littoral fauna of the North Atlantic most progressive, that of the North Pacific less so, the tropical faunæ containing many relict types of discontinuous distribution, and the Antarctic faunæ partly composed of types from the north which had crossed the barrier of warm water when the climatic zones were less differentiated than they now are; partly of groups developed in the south. Whether these groups were closely allied on the different southern continental shores would depend on their ability to cross the great barriers of deep ocean that separate them.

The distribution of the pinnipeds accords with these principles. The most specialized family is the Phocidæ, originating apparently in the Atlantic-Arctic basin, where *Phoca,* the most progressive genus, is found in the North Atlantic and Arctic seas and has penetrated into the North Pacific as far as California and Japan. Southward in the Atlantic it is succeeded by the less progressive *Monachus* in the Mediterranean and Antillean region. The Antarctic Phocidæ are also primitive and archaic, related more or less nearly to *Monachus.* In the Pliocene of Belgium are found extinct genera closely related to *Phoca* and others more primitive allied to *Monachus.*

The walruses, also Arctic and North Atlantic, have penetrated into the North Pacific only as far as Bering Sea; they are likewise recorded from the Pliocene of Northern Europe and along the North Atlantic in the Pleistocene as far south as Virginia.

The third family, the Otariidæ, is decidedly more primitive in structure, being less specialized for marine life. They are found in all the southern seas and on the North Pacific coasts. They are unknown to the North Atlantic and Arctic shores and have never been found fossil in either Europe or eastern North America. *Desmatophoca* and *Pontoleon* of the Miocene of Oregon are perhaps ancestral types, but more evidence is necessary before its North Pacific origin can be regarded as satisfactorily indicated.

Insectivora

Among the Insectivora we deal with a number of very ancient races, whose relationship is much more distant than in many other mammalian orders. They are small, and most of the surviving members are scarce and little known, while they are still less known as fossils. So far as we have any satisfactory evidence, the different races or most of them appear to have originated in the Holarctic region and spread to the southward. The most primitive division, the zalambdodonta, includes four families,

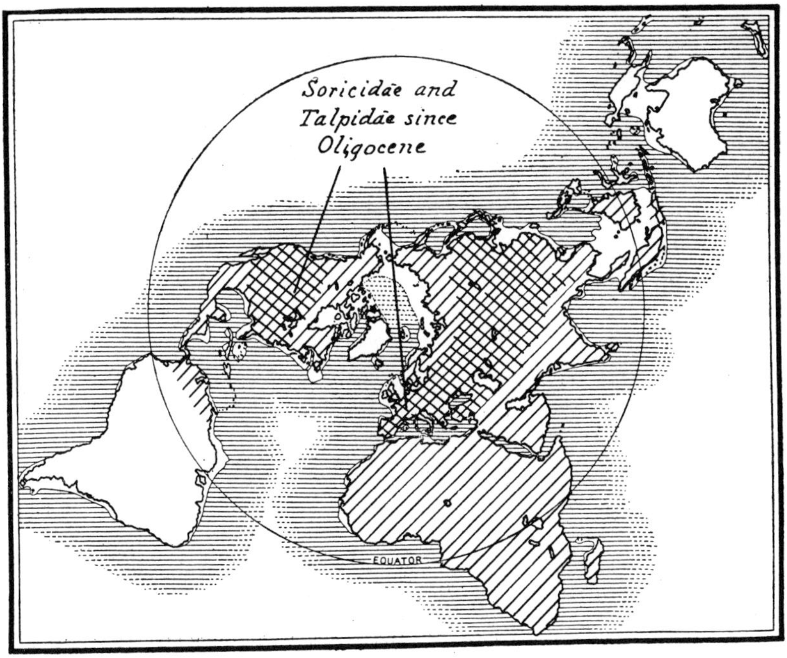

FIGURE 12. Distribution of Soricidæ (right to left shading) and of Talpidæ (left to right shading)

The less specialized Soricidæ are more widely dispersed, the highly specialized Talpidæ limited to Holarctica. Ancestral types of both are found in the Tertiary of Europe and North America, but the evidence as to their phylogeny is very scanty.

the Centetidæ of Madagascar, Chrysochloridæ of South Africa, Potamogalidæ of West Africa and Solenodontidæ of Cuba. Fossil zalambdodonts are found in the late Miocene in South America, in the early Oligocene (and recently in the Basal Eocene[42c]) in North America. These indications are in conformity with the derivation of the group from very ancient Holarctic ancestors, the modern zalambdodonts being the last surviving remnants of a dispersal from the north in early Tertiary or possibly pre-Tertiary times. But the evidence is too slight to be conclusive.

[42c. Paleocene.—E.H.C.]

The hedgehogs are more clearly of Palæarctic origin, the most primitive survivors being the East Indian *Gymnura* and *Hylomys,* while the most progressive genus, *Erinaceus,* is Palæarctic and is preceded in the Oligocene and Miocene of Europe by more primitive ancestral forms. A relatively primitive genus, *Proterix,* occurs in the Oligocene of South Dakota, as a contemporary with more progressive genera in the Oligocene of Europe. The family is otherwise unknown in the New World.

The moles and shrews are also evidently of northern origin. Of the two families, the Soricidæ are more primitive in structure and have spread more widely; the more specialized Talpidæ are still limited to Holarctica, and in the extreme north their exclusion from the areas of permanently frozen subsoil has split their range into two disconnected areas. The most progressive and abundant shrews are Holarctic, while the Oriental and African species (Crocidurinæ) retain some primitive characters. Fossil moles and shrews in the middle Tertiary of Europe and America indicate that the divergence between the two families was not then so great as now. The modern genera are reported to occur (but on inadequate evidence) as early as the Miocene in Europe and America. Jaws of several minute talpoid genera are known from the Middle and Lower Eocene of North America. They are unknown in the extra-Holarctic Tertiary, but this negative evidence is of no weight in view of their minute size and rarity.

The Tupaiidæ of the East Indies and Macroscelididæ of Africa occupy a somewhat anomalous position, since they are of higher type in brain development than other Insectivora and in many respects are nearer to the higher placental mammals.[43] Their distribution so remote from the great northern dispersal center may perhaps best be accounted for by considering the fact that their specializations, adaptations and habits of life are of a less unusual kind than in most of the lower insectivores and would bring them more directly into rivalry with certain groups of rodents, with which they were unable to contend successfully and were compelled to retreat southward in consequence. No fossil remains certainly referable to these families are known, although quite a number of early Tertiary genera of Europe and North America have been or might be provisionally referred to them.[44]

---

43. This anomaly in distribution is now removed by the studies of Gregory and Elliot Smith, which show that the true relations of *Tupaia* and presumably of *Macroscelides,* are with the Primates, rather than with the Insectivora. Their geographic distribution is quite normal on this view of their affinities.

44. *Entomolestes* of the Middle Eocene of North America is regarded by Dr. Gregory as probably related to *Tupaia,* and a number of other small mammals from the Bridger and Wasatch may be related to this group of Insectivora.

There are a large number of primitive Insectivora in the Eocene of North America and a few in Europe, which do not seem to be nearly ancestral to any modern group but rather indicate that the order once took a much more important place in the mammalian faunæ of the world than it does now. This should be kept in mind in considering the relations of the Insectivora.

Chiroptera

I am not sufficiently acquainted with modern Chiroptera to venture an opinion as to whether or not their geographical distribution indicates their place of origin, but I should not expect to find much satisfactory evidence, as they are known to be of very ancient specialization and to have greater facilities for wide distribution than terrestrial animals.

Dr. Andersen,[45] in his recent Catalogue of the Chiroptera in the British Museum, remarks: "The evidence afforded by the geographical distribution of Bats has generally been considered of doubtful value; hence they have either been entirely excluded from the material worked out by zoögeographers, or at least treated with pronounced suspicion as likely to be more or less unreliable documents of evidence. This unwillingness or hesitation to place Bats on an equal zoögeographic footing with non-flying Mammalia would seem to be due partly to the preconceived idea that owing to their power of flight Bats must evidently have been able easily to spread across barriers which in ordinary circumstances are insuperable for wingless Mammalia; partly to the fact that hitherto very often whole series of distinct forms have been concealed under one technical name. . . ." [the author cites a series of instances of this kind which] "tend to show that the present distribution of the Megachiroptera has not been influenced to any great, and as a rule not to any appreciable extent by their power of flight; if it had the Fruit-bat fauna of islands could not so commonly as is actually the case differ from that of a neighboring group or continent, and the tendency to differentiation of insular species or forms would have been neutralized by the free intercourse between neighboring faunas."

The belief that bats are more easily able to cross ocean barriers than non-flying mammals is probably based, not on the preconceived idea that they could, but upon the plain fact that they have done so far more frequently. Birds and bats are found upon numerous oceanic islands where

45. K. Andersen: Catalogue of the Chiroptera in the British Museum. 1: (Megachiroptera) lxxvi. 1912.

no non-flying mammals, and very few non-flying animals at all, exist. That they have wings and occasionally use them for so long a journey, whether voluntarily or involuntarily, is a natural explanation. I cannot see any other reasonable interpretation of the fact that they are present and the terrestrial mammals absent in so many remote oceanic islands. With bats, as with most birds, the intervening ocean acts as a hindrance, but their wider distribution shows that it is less of a hindrance than with terrestrial mammals.

((The transportation across oceanic or other barriers for most forms of life has been probably through agency of storms, whirlwinds and waterspouts. Normal winds are not a large factor, because most animals are adapted in structure and habits to resist or find shelter from these or any other customary incidents of life. But they are not able to resist the occasional violent fury of a storm, and all kinds of small objects may be lifted high in the air and carried hundreds of miles before they come to the ground again. Tornadoes and small rising whirls of wind, while more localized are very effective, and I have seen accounts of small ponds being completely emptied by such storms, their contents being carried high into the air, and transported probably for a long distance.

The various accounts of "rain of fishes" or of frogs, are due to those cases, necessarily very rare, where the material thus transported had such quantities of small animals in it as to attract notice. Transportation of a few animals or of a bunch of eggs on a leaf or bit of wood must be a rather common occurrence, so much so that the problem would rather be why if barriers are so easily crossed, do they seem to be so effective in limiting . . .)) [unfinished]

((The great antiquity of the order is indicated by its fossil record. The earliest bat is *Zanycteris* of the Upper Paleocene (Tiffany beds) of New Mexico, based upon an imperfect skull. It appears to be definitely related by the dentition to the Neotropical Phyllostomatidæ. *Provampyrus* Schlosser from the Lower Oligocene of Egypt is referred to the same family, but quite without evidence, as it is based on a wing-bone which gives no clue to family relationship. From the Upper Eocene and Oligocene of Europe numerous well-preserved remains of Chiroptera have been described, some *Archæonycteris, Palæochiropteryx* referred to extinct families of somewhat aberrant specialized type, others to the existing widely distributed Rhinolophidæ and Vespertilionidæ. The frugivorous bats, now widely distributed throughout the tropical Old World, are represented in the Oligocene of Italy. None of these show any markedly primitive stages in the evolution of the order. It is evident, on the contrary, that the differentiation of the Chiroptera from other primitive placentals must go back much earlier than the beginning of the Tertiary.))

Rodentia

The abundant and dominant order of Rodentia lends, in general, strong support to the theories here advocated; but there are certain serious difficulties which can be reconciled only by appealing to the imperfection of the geological record.

The rabbits and picas form a group apart, the former Nearctic, the latter Palæarctic since the Oligocene, and both Holarctic since the Pleistocene, the rabbits having extended their range over most of the Oriental region and a large part of the Ethiopian and Neotropical. A single specimen is recorded as from the Pleistocene of South America; their introduction to Australia is known to have been by civilized man.

Of the remaining rodents, the myomorph families are evidently of Holarctic origin, as they first appear in Europe and North America in the Oligocene and the highest and most progressive modern types (*e. g.*, Arvicolinæ) are now Holarctic, while in the southern continents they are unknown until the Pleistocene and various primitive survivals are found still living in Oriental, Ethiopian and Neotropical regions. We may note, however, that the very abundant and typical group of Cricetinæ has its most primitive living representatives in tropical regions, that as we go south in South America, the genera approximate more towards the more specialized arvicoline type, in the same way that they do as we go northward in the northern continents.[46] Since there is no doubt that the cricetines are of northern origin, appearing first in South America in the Pliocene or Pleistocene, while they are common in the Holarctic regions from the Oligocene to the present day, we must suppose that the higher development of the Antarctic genera, to which Oldfield Thomas has called attention, is a case of parallelism with that of the Arctic genera and that the colder climate of the far south is the stimulus which reversed the usual conditions of geographical distribution. A review of the fauna of the Argentine as compared with that of tropical South America tends to show, I think, that this condition is general throughout, and that the fauna is more progressive and more nearly equivalent in development to those of the northern world than is that of the intervening tropical zone. This is equally true of autochthonic races and of those which are demon-

---

46. OLDFIELD THOMAS.

strably of northern origin. Compare distribution of the genera of Procyonidæ, Canidæ, Cervidæ, Tatuidæ and Dasypodidæ among mammals.

Among the sciuromorphs, the squirrels are of early appearance (Oligocene) in the northern world but are now most abundant in the East Indies. The more specialized and later appearing marmots are chiefly Holarctic. The highly specialized beavers and pocket-gophers are Holarctic and Nearctic respectively, from their first appearance. A marked excep-

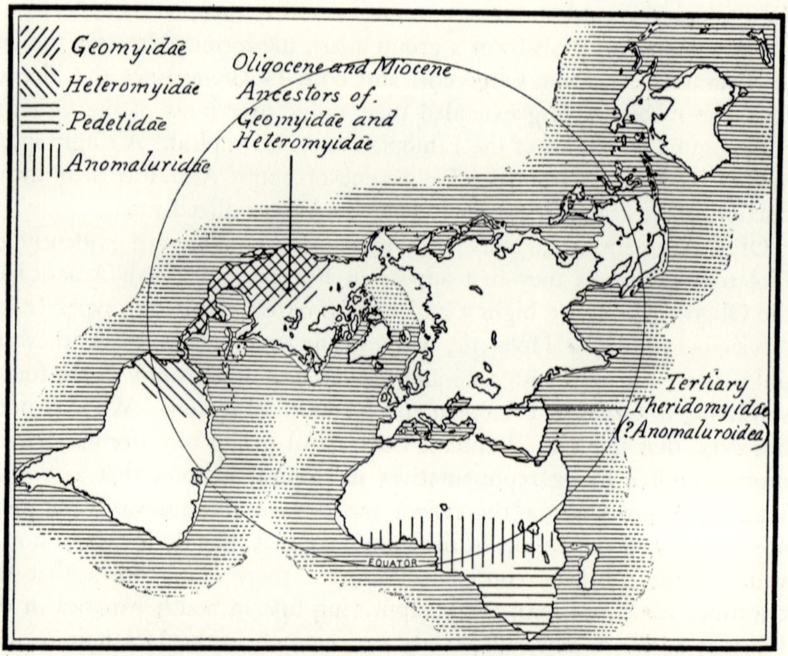

FIGURE 13. Distribution of Geomyoidea, Anomaluridæ and Pedetidæ

The Geomyoidea are of Nearctic origin, but the more primitive Heteromyidæ have spread into part of South America. The Anomaluridæ are thought to be the nearest living relatives of the early Tertiary Theridomyidæ. The Pedetidæ are an aberrant specialization, derived perhaps from the same group.

tion to the rule is seen in the survival in the western Sonoran subregion of *Aplodontia,* the most primitive living sciuromorph in several respects. I have no explanation to offer of this anomaly, save that we have not yet balanced properly the essential qualities of progressiveness among Rodentia.

Among the hystricomorphs, we find serious difficulties in the distribution. The most primitive living group is certainly the Anomaluridæ of

West Africa; but, like the Pedetidæ of South Africa, they offer a puzzling admixture of characters, which makes it doubtful whether they should be reckoned as pertaining to the same stock as the other hystricomorphs. The remaining families, while chiefly South American, are also partly represented in the Ethiopian, Oriental and Holarctic regions. It may be possible, in view of the facts that the European Theridomyidæ antedate

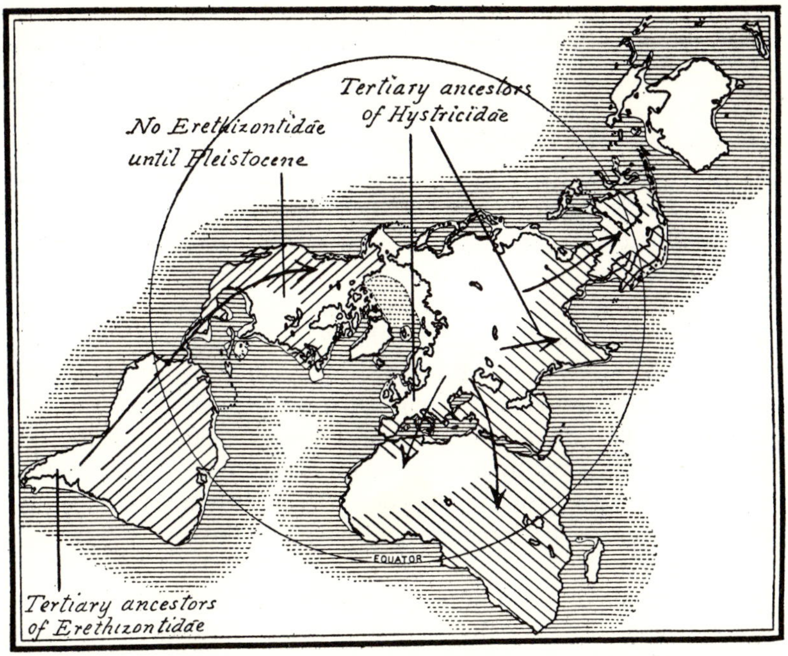

FIGURE 14. Distribution of the true porcupines (Hystricidæ) and New World porcupines (Erethizontidæ)

The Hystricidæ appear to be of Palæarctic dispersal, the Erethizontidæ are apparently of Neotropical origin.

geologically any specialized hystricomorphs, are apparently directly intermediate between the primitive rodent type (*Paramys* and its allies) and the hystricomorphs and show the early stages of differentiation of several hystricomorph families, that the Hystricomorpha are a group of Holarctic origin which has spread into all the southern continents and specialized independently on parallel lines. But their entire absence from the recorded North American Tertiary is then explainable only by the defective record, and our knowledge of North American Tertiary rodents is so extensive that I should hardly regard this assumption as justifiable. The fact that the highest and most specialized types are South American necessarily involves the idea that that continent has been the most important center of their later development and dispersal, and the alliance of the African to the South American genera and of the New and Old World porcupines must be regarded as more remote than it appears. Dispersal from South America by help of Antarctic or transatlantic land bridges will not solve the problems of their distribution much better. The most specialized porcupines in most respects are the hystricids of the Old World —late Tertiary in Europe, now chiefly Oriental and African. The Nearctic porcupines (*Erethizon*) are more advanced in several features than the Neotropical (*Synetheres*). Yet the ancestors of the New World porcupines at least occur in the late Tertiary of South America and are absent or unrecorded from the Tertiary of North America. The distribution of the Octodontidæ in Africa and South America would possibly admit of being interpreted by parallel development from theridomyid ancestors; but the parallelism must have been singularly close, and the absence or non-recognition of Theridomyidæ from the North American Tertiaries appears surprising. I have been unable to frame any hypothesis which will fit all the facts of distribution in this group,[47] except by assuming that the South American Hystricomorpha, which as Scott has shown are all clearly derived from a single stock, reached South America from Africa in the Oligocene by over-sea raft transportation. This involves so long a voyage that I hesitate to accept it as a reasonable probability, even though the winds and currents would obviously favor transportation in this direction.

The Hystricidæ may fairly be assumed as of Old World origin, and probably Palæarctic, since they are represented in the later Tertiary of Europe and are unknown in the New World. The Erethizontidæ must apparently be derived from South America, since they are unknown in the Old World, and unknown in the North American Tertiary, while *Steiromys* of the Patagonian Miocene appears to be ancestral.

The primary type of the simplicidentate rodents, as I have elsewhere shown,[48] must be regarded as being represented by the Ischyromyidæ of the American and European Eocene, in particular by *Paramys* and *Sciu-*

47. The hypothesis of migration to or from South America across a land bridge from Africa to Brazil is equally unsatisfactory as an attempt to explain the relations of the hystricomorph families and is entirely at variance with the evolution and distribution of other mammalian orders, besides being highly improbable on isostatic grounds. The supposed evidence in its favor from lower vertebrates and invertebrates is due, so far as I have been able to examine it, to a lack of appreciation of the principles of dispersal of races and of parallelism and of the imperfection of the geological record.

48. "Osteology and Relationship of Paramys and Affinities of the Ischyromyidæ," Bull. Am. Mus. Nat. Hist. **28**: 43-71. 1910.

ravus. All other rodents may be derived from this group by divergent, parallel and in some respects convergent evolution. Modern rodents represent a great number of independent derivations from this primary stock, their association into sections and families being to a considerable extent artificial.

There are no rodents in the *Notostylops* Beds of South America (Eocene); presumably therefore none in preceding epochs. There are none in the Paleocene of Europe and North America; presumably therefore their sudden appearance in the true Eocene of these regions was due to migration from some other region, equidistant from either, as their development is almost equivalent in the two,—therefore probably Asia. The few Theridomyidæ of the Oligocene of Africa are rather primitive forms, certainly not more progressive than their contemporary relatives in Europe, affording thus a slight indication that they were Palæarctic immi-

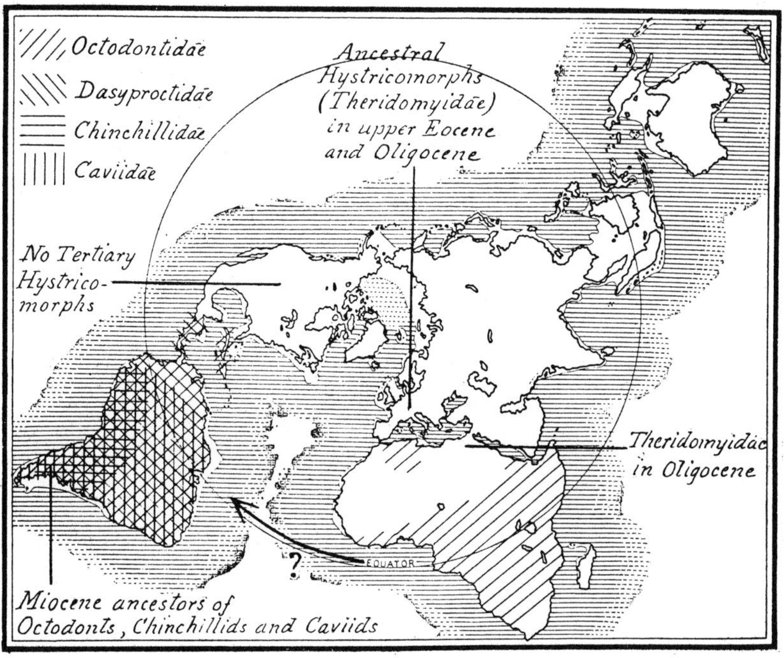

FIGURE 15. Distribution of the Neotropical families of Hystricomorphs

The Octodontidæ are also found in Africa, and the Theridomyidæ of the early Tertiary of Europe are apparently ancestral to these families of the Hystricomorpha. No hypothesis satisfactorily explains the accepted relationship and distribution.

TABLE 4

DISTRIBUTION OF THE RODENTS

	S. America	N. America	Asia	Europe	Africa	Australia
Recent	Muridæ and Hystrico-morpha Lagomor-pha	Myomorpha Sciuromor-pha Lagomor-pha *Erethizon*	Myomorpha Sciuromor-pha Lagomor-pha Hystricidæ (Orie.)	Myomorpha Sciuromor-pha Lagomor-pha	Myomorpha Sciuromor-pha Hystrico-morpha	Muridæ
Pleisto-cene	Hystrico-morpha and Muridæ Lagomor-pha	Myomorpha Sciuromor-pha Lagomor-pha *Erethizon*	Myomorpha Sciuromor-pha Lagomor-pha Hystricidæ	Myomorpha Sciuromor-pha Lagomor-pha Hystricidæ	?	?
Pliocene	Hystrico-morpha	Sciuromor-pha Myomorpha Leporidæ	Myomorpha Leporidæ? Hystricidæ	Myomorpha Sciuromor-pha Ochotonidæ Hystricidæ	?	
Miocene	Hystrico-morpha	Sciuromor-pha Myomorpha Leporidæ	?	Sciuromor-pha Myomorpha Ochotonidæ	((Pedetidæ))	
Oligocene	Cephalo-mys49	Sciuromor-pha Myomorpha Leporidæ Ischyromy-idæ	((Criceti-dæ)) ((Bathyer-gidæ)) ((Ochotoni-dæ))	Sciuromor-pha Myomorpha Ochotonidæ Therido-myidæ	Therido-myidæ	
Eocene	None	Ischyro-myidæ	((Ochotoni-dæ))	Therido-myidæ Ischyro-myidæ	?	
Paleocene		None	((*Baënomys*))	None		
Creta-ceous		None	?	None		

grants. In Australia the evolution of Marsupial analogues of the more abundant rodent types of Arctogæa affords strong evidence that the true rodents were absent from Notogæa until the end of the Tertiary; a view

49. A hystricomorph, recorded by Ameghino from the *Pyrotherium* beds.

confirmed by the limited amount of adaptive radiation which the invading Muridæ have undergone in that continent up to the present day.

The Australian Muridæ can only be accounted for by over-sea transportation, for the family appeared and evolved during the middle and later Tertiary, and the peculiarities of the Australian fauna are explained by all writers as due to isolation extending through the Tertiary period.

Perissodactyla

The order Perissodactyla is represented to-day by three widely separated families—the rhinoceroses, Ethiopian and Oriental; the tapirs, Neotropical and Oriental, and the horses, Asiatic and Ethiopian. The last group is the most progressive and modernized, but the whole order must be regarded as having seen its best days and as passing towards extinction in competition with the better organized and more adaptable Artiodactyla.

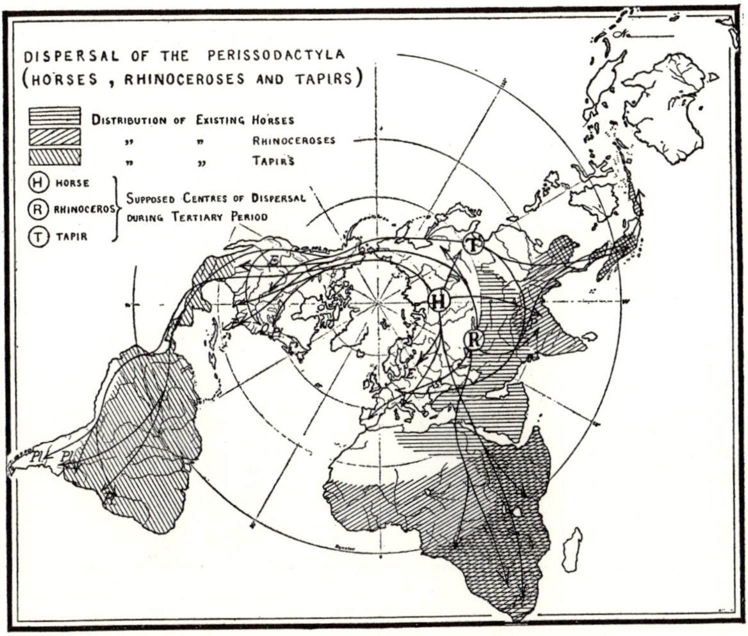

FIGURE 16. Dispersal and distribution of the Perissodactyla

The tapirs are on the whole the most primitive and their present distribution widely discontinuous. The rhinoceroses are less widely dispersed and the horses the most central in their present distribution. All were inhabitants of Tertiary Holarctica, but their dispersal centers appear to have been Palæarctic, as indicated.

The geological record affords abundant evidence of the Holarctic origin of all the Perissodactyla. The ancestry of each race can be traced back in the Tertiary faunæ of Europe and the United States, in a series of approximately ancestral stages, sometimes closer in one region, sometimes in the other, to a group of closely allied primitive perissodactyls in the early Eocene of both countries. In South America, the order is unknown until the late Pliocene and Pleistocene. In other regions we know too little of the early Tertiary faunæ to say when the perissodactyls first appeared, but they are absent from the Oligocene fauna of Egypt, from the Pleistocene and modern faunæ of Australia and of all oceanic islands. This accords with the natural inference from their size, proportions and habits that they would be strictly limited by land connection in their geographic distribution.

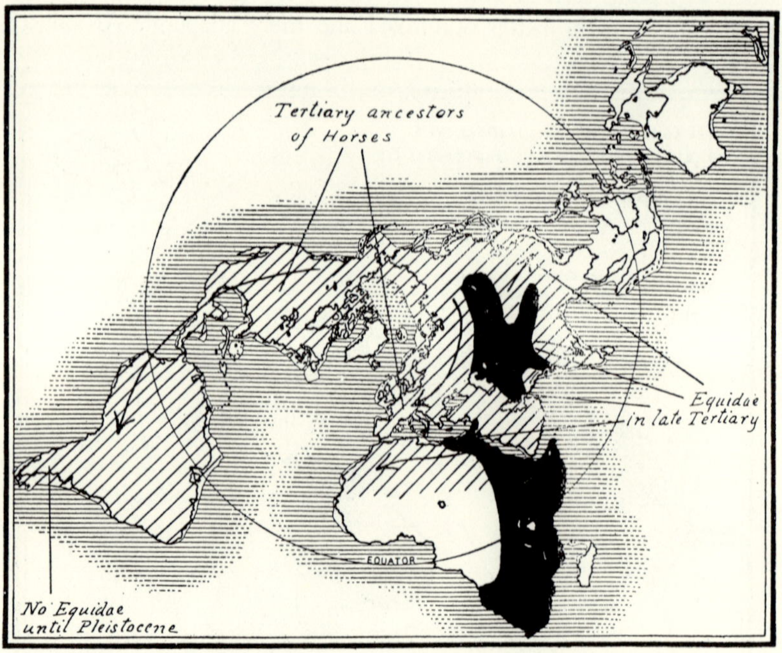

FIGURE 17. Distribution of Equidæ, living (solid black) and Pleistocene (shaded)

Early Tertiary ancestors are found in Nearctic and Palæarctic regions. The American series is more direct than that of Europe until the late Tertiary. This and other considerations indicate the center of dispersal as in northeastern Asia or northwestern North America.

Besides the surviving groups, the early perissodactyls gave rise to several extinct families, the lophiodonts, palæotheres, titanotheres and chalicotheres, none of which are known to have invaded the southern continents.

Equidæ.—The best known phylum of the order, that of the horses, is certainly not a direct genetic succession, as regards known species, but approximately so as regards the known genera. The successive genera are progressively more specialized in accordance with their geological sequence. They are identical or closely allied in the European and North American sequence. In North America, the series is more complete, the

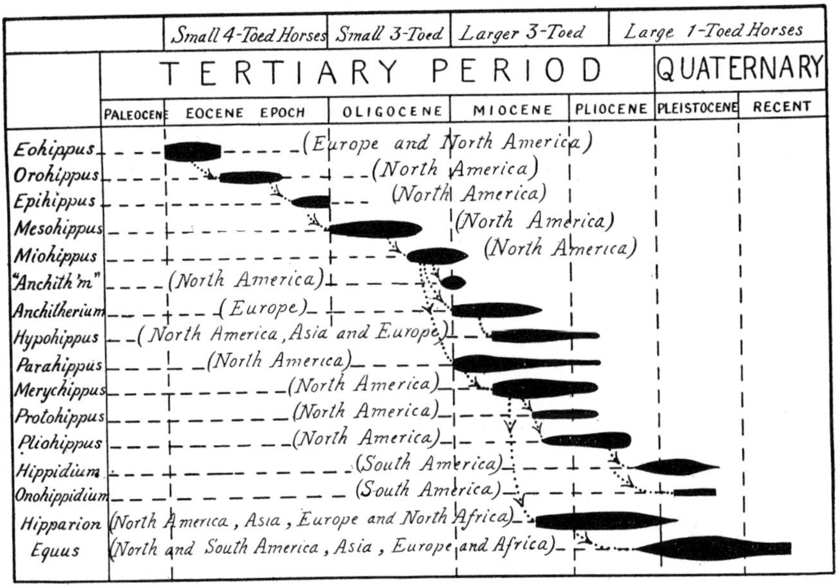

Figure 18. Geologic range and phylogenetic relations of fossil Equidæ

The overlap in geologic range of the genera, and the sudden appearance of each new stage, indicate that our record is not derived from the center of dispersal of the race; although the American series is sufficiently direct to indicate that it was not very remote.

approximation to a direct genetic sequence is closer and the successive stages appear earlier in time. This is reasonably interpreted by supposing that the center of dispersal was intermediate between Europe and the western United States but nearer to the latter.[49a] That is to say, it was either in boreal North America or in northeastern Asia. The absence of primitive survivors of the race in the East Indies is natural; as the horses were very early adapted to open plains, unfitted for mountain or forest habitat, the great transverse Himalayan chain would form an almost im-

[49a. Recent work has tended to confirm the view that this center was in North America.—E.H.C.]

passable barrier and the heavily forested regions of the East Indies would have no attractions to tempt the ancestral horses to pass around its eastern end.

TABLE 5

DISTRIBUTION OF THE EQUIDÆ

	Neotropical	Nearctic	Palæarctic	Oriental	Ethiopian
Recent			*Equus*	*Equus*	*Equus*
Pleistocene	*Equus* *Hippidion* *Onohippidion*, etc.	*Equus*	*Equus*	*Equus* (India)	*Equus*
Pliocene	——	*Hipparion* *Pliohippus* *Protohippus* *Hypohippus*	*Equus* *Hipparion* *Hypohippus*	? *Equus*[49b] (Siwalik) *Hipparion*	
Miocene	——	{ *Merychippus* { *Hypohippus* { *Merychippus* { *Hypohippus* { *Parahippus* *Parahippus*	*Hipparion* *Anchitherium* *Anchitherium*	*Hipparion* (L'r Siwalik)	
Oligocene	——	*Miohippus* *Mesohippus*			——
Eocene	——	*Epihippus* *Orohippus* *Eohippus*	{ *Anchilophus* { *Lophiotherium* *Pachynolophus* *Hyracotherium*		

TAPIRIDÆ.—The tapirs are the most primitive living perissodactyls, retaining the primitive number of digits in fore and hind feet and the primitive short-crowned grinding teeth. They are to-day limited to the East Indies and tropical America. In the Pleistocene, they inhabited the Sonoran region and continental India and the marginal parts of the Palæarctic region. Their Tertiary ancestry has been traced back in Europe and in

[49b. *Equus* is definitely established as having appeared for the first time in India in the Pleistocene.—E.H.C.]

TABLE 6

DISTRIBUTION OF THE TAPIRIDÆ

	Neotropical	Nearctic	Palæarctic	Oriental
Recent	*Tapirus* s. s. *Tapirella*	None	None	*Tapirus* (*Rhinochœrus*)
Pleistocene	*Tapirus* (s. l.)	*Tapirus* (s. l.)	*Tapirus* (s. l.)	*Tapirus* (*Rhinochœrus*)
Pliocene	None	?	*Tapirus* (s. l.)	
Miocene	None	"*Tapiravus*"	*Tapirus* (s. l.) *Paratapirus*	?
Oligocene	None	*Protapirus*	*Protapirus*	?
Eocene	None	*Isectolophus* *Helaletes*[50] *Systemodon*	*Lophiodontidæ*[51]	?

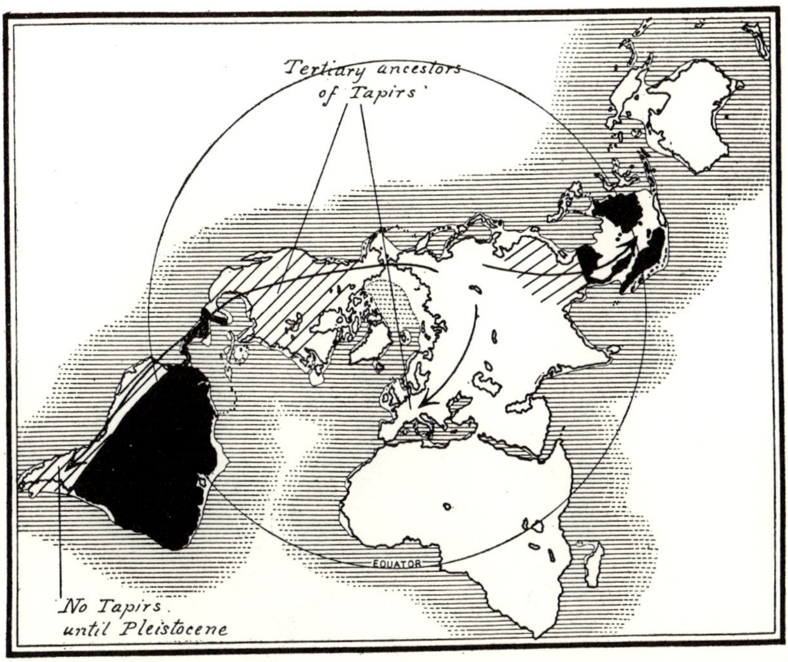

FIGURE 19. Distribution of the Tapirs living (solid black) and Pleistocene (shaded)

Ancestral types are found in the Tertiary formations of Europe and North America. The relations of the two series and the Pleistocene and modern distribution indicate a dispersal center in eastern Asia.

50. True affinities of these genera require revision.
51. Affinities between tapirs and rhinoceroses.

North America to the Oligocene *Protapirus,* which is preceded by a less direct ancestral series in the Eocene of North America; but ancestral tapirs have not been identified in the Eocene of Europe. The data are insufficient to determine the center of dispersal except as probably in the Palæarctic region. Tapirs are unknown in South America until the Pampean (Pleistocene); they do not appear to have reached Africa at all. The arid climate of the Afro-Asiatic connection and the heavily forested path of migration to the East Indies would seem to be the features that determined the dispersal of the horses into Africa, the tapirs into Malaysia.

Rhinocerotidæ.—The rhinoceroses are intermediate between horses and tapirs in adaptation. The Tertiary history of the group is much the same, approximate series being found in Europe and North America as far back as the Oligocene or Eocene, but the phyla are less direct and complete, and there is a greater diversity of type among them. The Palæarctic series appear to be more direct, and this, in connection with the fact that the

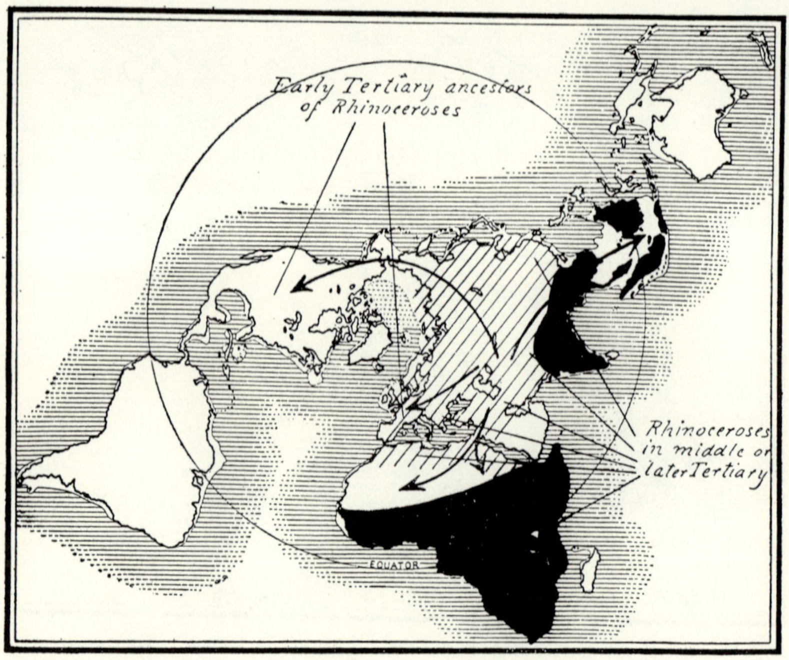

Figure 20. Distribution of the Rhinoceroses, living (solid black) and Pleistocene (shaded)

Primitive rhinoceroses are found in the Palæarctic and Nearctic Tertiaries and late Tertiary of India and Africa. Comparison of the Palæarctic and Nearctic series indicates that the center of dispersal was in west-central or southwestern Asia.

race never reached South America, may be taken to indicate that the center of dispersal was Palæarctic rather than Nearctic, less northerly than that of the horses, less easterly than that of the tapirs. At all events, the relations of the later Tertiary rhinoceroses indicate that North America was much more remote from the center of dispersal than Europe, while southwestern Asia was very close to it.

Table 7

Distribution of Rhinoceroses

<table>
<tr><th></th><th>Nearctic</th><th>Palæarctic</th><th>Oriental</th><th>Ethiopian</th></tr>
<tr><td>Recent</td><td>None</td><td>None</td><td>Rhinoceros
Ceratorhinus</td><td>Cœlodonta
Opsiceros</td></tr>
<tr><td>Pleistocene</td><td>None</td><td>Elasmotherium
Cœlodonta
Opsiceros
Ceratorhinus</td><td>Rhinoceros
Opsiceros</td><td>Opsiceros</td></tr>
<tr><td>Pliocene</td><td>Teleoceras</td><td>Opsiceros
Ceratorhinus
Teleoceras</td><td>Rhinoceros
Teleoceras
Aceratherium</td><td>?</td></tr>
<tr><td>Miocene</td><td>Teleoceras
Aphelops
Diceratherium</td><td>Teleoceras
Aceratherium
Diceratherium</td><td>Rhinoceroses</td><td>Rhinoceroses</td></tr>
<tr><td>Oligocene</td><td>Cœnopus
Trigonias
Metamynodon
Hyracodon</td><td>Cœnopus[52]
Prohyracodon</td><td>Aceratherium[54]
Diceratherium
Teleoceras
Cadurcotherium</td><td>No Perissodactyla</td></tr>
<tr><td rowspan="2">Eocene</td><td>Amyndon
Triplopus</td><td></td><td rowspan="2">?</td><td rowspan="2"></td></tr>
<tr><td colspan="2">Lophiodontidæ[53]</td></tr>
</table>

Artiodactyla

The great and diverse order of Artiodactyla can fairly be regarded as of Holarctic origin as a whole. Its distribution can most readily be considered group by group.

Pigs and Peccaries.—These two groups are characteristic of the Old and New World respectively. The pigs are now chiefly Ethiopian and

52. Includes a number of subgenera recently defined by Abel.

53. This family may be regarded as ancestral to both rhinoceroses and tapirs, but the more exact derivation is doubtful.

54. Gaj fauna, upper Aquitanian auct. Pilgrim. It should perhaps be regarded as Lower Miocene.

Oriental, the peccaries Neotropical in distribution. The peccaries first reached South America in the Pleistocene and ranged throughout the United States from the Oligocene to as late as the Pleistocene. Pigs were common in the Oligocene and later Tertiary in Europe and were present in India in the Miocene, probably earlier. The Tertiary ancestry of the pigs in Europe can be traced back to a common ancestral group in the Eocene, and the same is true of the peccaries in the western United States.

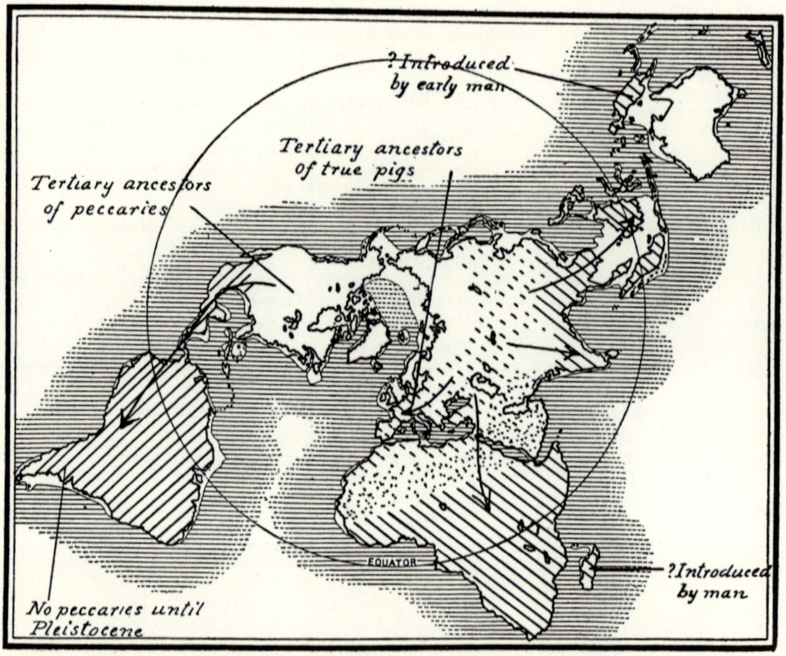

FIGURE 21. Distribution of pigs and peccaries

In Old World, broken shading *Sus* only; full shading, other genera. In New World, full shading *Dicotyles*. The dispersal center of Dicotylidæ was Nearctic, of Suidæ Palæarctic. The living South American genus is more primitive than the Pleistocene genera of North America, *Platygonus* and *Mylohyus* (the Pleistocene North American species referred to *Dicotyles* are all *Mylohyus*).

RUMINANTS.—Under this term, we may conveniently include all the selenodont artiodactyls,—the camels and tragulines, deer, antelopes, sheep and cattle, besides various extinct groups.

They are admittedly of northern origin. In South America, they do not appear until the end of the Tertiary (*Microtragulus,* Monte Hermoso); their representatives in the Oligocene of North Africa are much

TABLE 8

DISTRIBUTION OF THE PIGS AND PECCARIES

	Neotropical	Nearctic	Palæarctic	Ethiopian	Oriental
Recent	*Dicotyles*		*Sus*	*Potamochœrus* *Phacochœrus*	*Sus* *Babirussa*
Pleistocene	*Dicotyles*	? *Dicotyles* *Mylohyus* *Platygonus*	*Sus*		
Pliocene	? *Platygonus* ? *Prosthennops*	*Platygonus*	*Sus*		*Sus, Hippohyus* *Samitherium* *Potamochœrus*
Miocene	None	*Prosthennops* *Desmathyus*	*Sus* *Listriodon* etc.	(No record)	*Listriodon* ? *Sus, Hippohyus* *Potamochœrus* *Hyotherium* *Palæochœrus*
Oligocene	None	*Perchœrus*	*Palæochœrus* *Hyotherium* etc.	None[55]	*Palæochœrus*[56]
Eocene	None	? *Helohyus* etc.	*Cebochœrus* etc.	(No record)	

more primitive than the contemporary artiodactyls of Europe; the highest and most progressive types are found to-day in Asia, and the most antique and primitive survivals in the East Indies, West Africa and tropical America. The several groups indicate in their present distribution, and in what is known of their past history, that their centers of dispersal were in different parts of the northern world, as we have seen among the perissodactyl groups.

The camels appear to have been of American origin. An ancestral series is found in the Tertiary of the western United States, going as far back as the Upper Eocene.[57] In the Old World, they first appear in the Pliocene; in South America, in the Pleistocene (Pampean); and the camels

55. Schlosser has shown that *Geniohyus* is a hyracoid, not an artiodactyl.

56. Gaj fauna, regarded by Pilgrim as upper Aquitanian.

57. It forms a singularly direct and complete phylum, so supercharged with intermediate and connecting forms that it is very difficult to classify and arrange the fossils into species and genera, while every gradation of structural evolution is abundantly illustrated.

of the Pleistocene in North America were about as specialized on the whole as the living llamas of South America or the camels of Africa and Asia. In North America, the race is now extinct. The center of dispersal would appear to have been in this continent,—how far to the north we have no means of estimating; but the exceptional directness of the phylogenetic series as represented by our western fossils indicates, in my opinion, that these fossils lived in or close to the racial dispersal center.

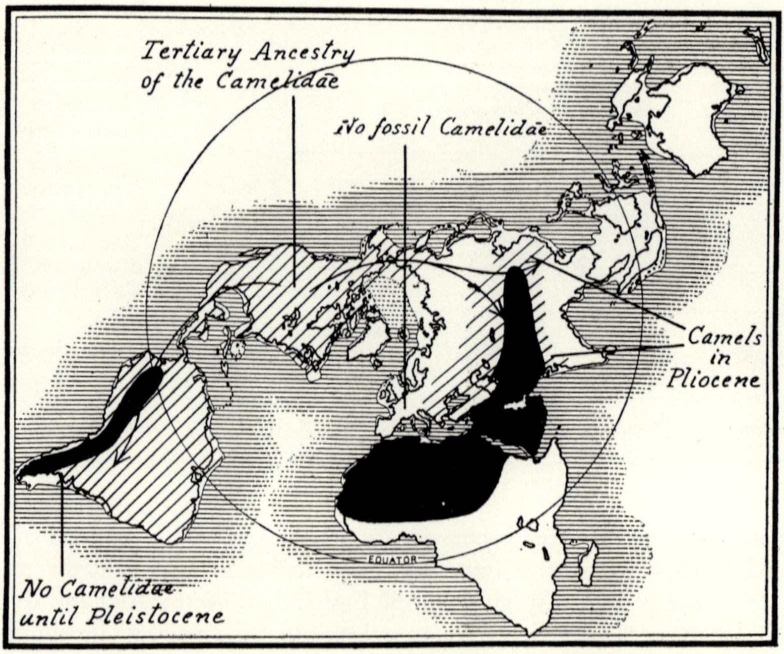

FIGURE 22. The dispersal center of the Camelidæ was in North America

They reached the Old World in the Pliocene, South America in the Pleistocene. They survive on the margins of their range but became extinct in North America early in the Pleistocene. North American Pleistocene camels were more advanced than the living types of the marginal areas.

The tragulines, recent and extinct, are a heterogeneous assemblage of primitive ruminants, whose real affinities have been much disputed. In the present writer's opinion, the living East Indian chevrotains should be associated with *Hypertragulus* of the North American Oligocene and perhaps *Microtragulus* of the South American Pliocene, and the center of distribution of this group hypothetically placed along the northeastern

TABLE 9

DISTRIBUTION OF THE CAMELS

	Neotropical	Nearctic	Palæarctic	Ethiopian	Oriental
Recent	*Auchenia*	None	*Camelus*	None	None
Pleistocene	*Auchenia*	*Eschatius* *Camelops* *Camelus*	*Camelus* ? *Procamelus*[58]	?	*Camelus*
Pliocene	None	*Pliauchenia,* etc.	? *Procamelus*[59] ((*Pliauchenia*))	(No record)	? *Camelus*[58]
Miocene	None	((*Pliauchenia*)) *Procamelus* *Protolabis* *Miolabis,* etc. *Oxydactylus*	None		None
Oligocene	None	*Protomeryx,* etc. *Poëbrotherium* *Eotylopus*	None	None	(No record)
Eocene	None	*Protylopus*	None	(No record)	

coast region of Asia (*cf.* tapirs). The living water-chevrotain (*Hyæmoschus*) and most of the so-called tragulines of the European Oligocene and Upper Eocene are to be regarded as primitive stages of true Pecora. *Leptomeryx, Protoceras* and *Heteromeryx* are related forms from the North American Oligocene. Among these primitive forms, some (*Leptomeryx*) display affinities to the deer, others (*Protoceras, Heteromeryx*) to giraffes and antelopes.

In the later Oligocene of Europe and the Miocene of the United States appear more definitely deer-like types (*Dremotherium, Blastomeryx*), and in the succeeding formations we find progressively higher types of deer in Europe and North America, but always appearing earlier in the Old World. The deer—excepting the isolated primitive survival represented by the "water-chevrotain," closely related to *Dorcatherium,* a Miocene genus in Europe—have not reached the Ethiopian region, but were easily

58. *C. sivalensis* of the Siwalik beds is doubtfully congeneric with the modern species and, along with the so-called *Procamelus* described by Mme. Pavlow from the Pleistocene of Russia, appears to be an intermediate stage between *Procamelus* and *Camelus.* [*Camelus* first appeared in the Oriental region in the Pleistocene.—E.H.C.]

59. A doubtful Camelid, based on a single upper molar from the Pliocene (or Miocene) of China.

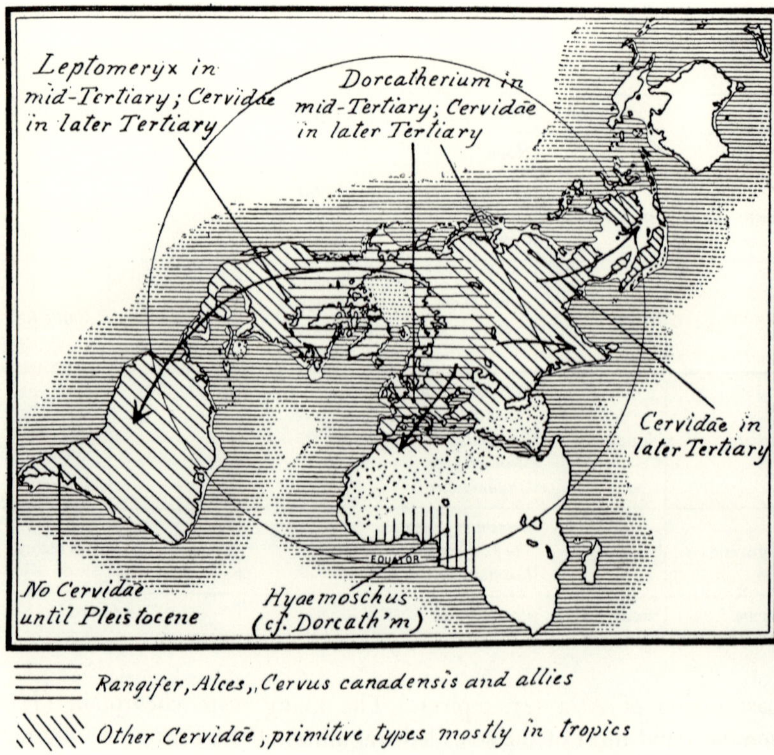

FIGURE 23. Distribution of Cervidæ and pro-Cervid Tragulidæ

The highest and latest appearing types are still confined to the circum-Arctic regions; the genera of the more peripheral regions are more primitive. The earliest and most direct ancestral series is found in Europe and Asia; the parallel series in North America is less direct and more retarded. A primitive survival is found in West Africa, protected by the desert from competition of higher types.

able to reach North America in the Pleistocene. I take it, therefore, that their center of dispersal was well to the east and north in Asia (*cf.* horses). Their migration into the Ethiopian region was checked after the Miocene by the progressive aridity of the desert region between, which served as a barrier to these forest-living ruminants, although not to the plains-living antelopes.

The antelopes, on the other hand, while also appearing fairly early in the European geologic record and abundant and well advanced in south-west and southern Asia as early as that record is revealed to our eyes, are imperfectly represented in North America—first appearing in the Pliocene and not widely varied even to-day, while they have not reached South America at all. They are to-day most abundant and varied in Africa.

TABLE 10

DISTRIBUTION OF TRAGULIDÆ PROPER

	Neotropical	Nearctic	Palæarctic	Ethiopian	Oriental
Recent	None	None	None	None	*Tragulus*
Pleistocene	None	None	None	None	*Tragulus*
Pliocene	? ? *Microtragulus*	None	None		*Tragulus*
Miocene	None	None	? None		None
Oligocene	None	[*Hypertragulus*]	? None	None	None
Eocene	None	[Primitive Artiodactyla]			

TABLE 11

DISTRIBUTION OF CERVIDÆ AND PRO-CERVID TRAGULINES

	Neotropical	Nearctic	Palæarctic	Ethiopian	Oriental
Recent	*Odocoileus* *Mazama*	*Cervus*, *Alces* *Rangifer* *Odocoileus*	*Cervus*, *Alces* *Rangifer*, *Dama*	*Hyæmoschus* (W. Africa)[61]	*Cervus* (*sensu lato*)
Pleistocene	*Odocoileus* *Mazama*	*Cervus*, *Alces* *Rangifer* *Odocoileus*	*Cervus*, *Alces* *Rangifer*, *Dama* *Megaceros*		*Cervus* (s. l.)
Pliocene	None	*Cervus* (s. l.)	*Cervus* (s. l.), etc.	(No record)	*Cervus* (s. l.) *Moschus* *Dorcatherium*
Miocene	None	*Dromomeryx* *Blastomeryx*	*Dorcatherium* *Dremotherium*	(No record)	*Dorcatherium*[62]
Oligocene	None	*Leptomeryx*,[60] etc.	*Prodremotherium* *Gelocus*, etc.	None	*Prodremotherium* *Gelocus*, etc.
Eocene	None	Primitive Artiodactyla ((*Archæomeryx*))		No record	

60. Family Hypertragulidæ, but *Leptomeryx* is structurally ancestral to American Cervidæ.

61. This group is referred generally to the Tragulidæ, but the common characters are persistent primitive features, and I regard it as a little altered survivor of the ancestors of the Cervidæ. Tragulidæ as here limited are a distinct phylum, primitive in many features but aberrant in others.

62. Family Tragulidæ as usually referred, but affinities are with *Hyæmoschus*, not with *Tragulus*; the group may fairly be regarded as ancestral to the Cervidæ, while the traguline group certainly is not.

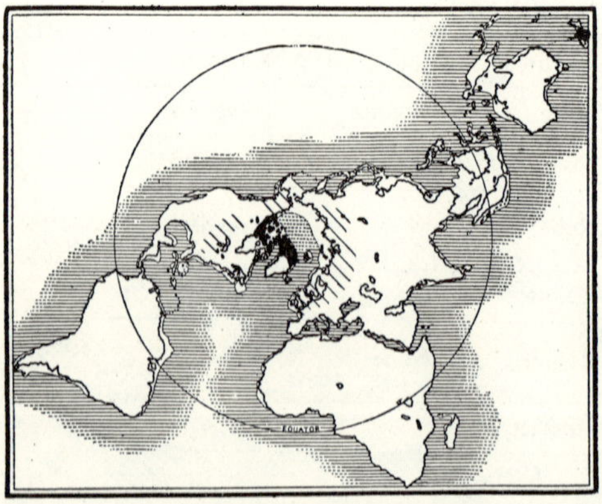

A. Muskoxen (Ovibovinæ)

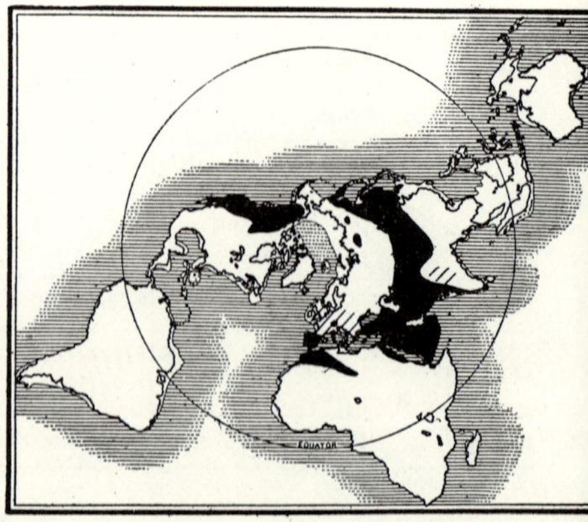

B. Sheep and goats (Ovinæ)

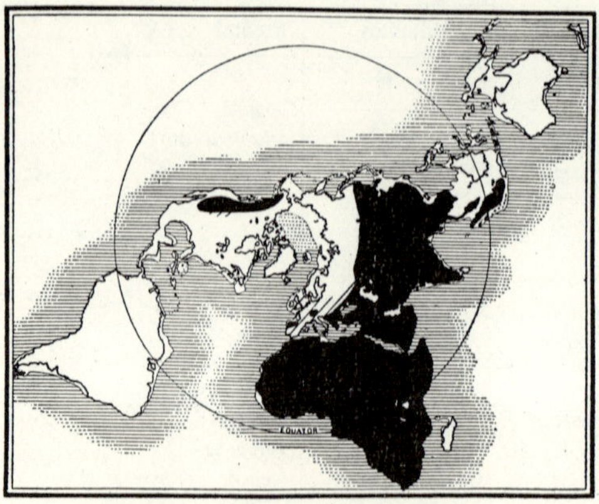

C. Other Antelopes (Antelopinæ etc.)

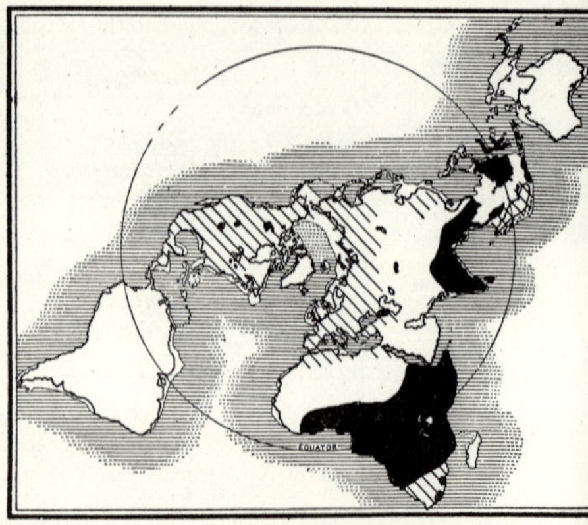

D. Cattle (Bovinæ)

FIGURE 24. Distribution of the Bovidæ, existing (solid black) and extinct (shaded)

The sheep and goats are regarded as the highest group; the muskoxen represent a specialized Arctic adaptation (*cf.* Eskimo among mankind). The cattle are a somewhat southerly type; their formerly wide northern distribution has been greatly restricted, and for the theory that they are of Oriental origin there does not appear to be any real evidence. The remaining Bovid subfamilies, usually grouped under the term "antelopes," are to a varying extent primitive and aberrant. The Holarctic groups are nearer to the sheep and goats and the more primitive groups are limited to the Ethiopian region and the East Indies.

From these facts, I infer that their center of dispersal was well to the west and south in Asia (*cf.* rhinoceroses). The sheep and goats are a comparatively recent development of the highest antelopes and must be assigned a center of dispersal somewhat more to the north.

The cattle are of comparatively recent appearance in Europe, as also in America. Judging from their present distribution, one would say that their center of dispersal was in southeastern Asia, the southward slopes of the Himalayas.

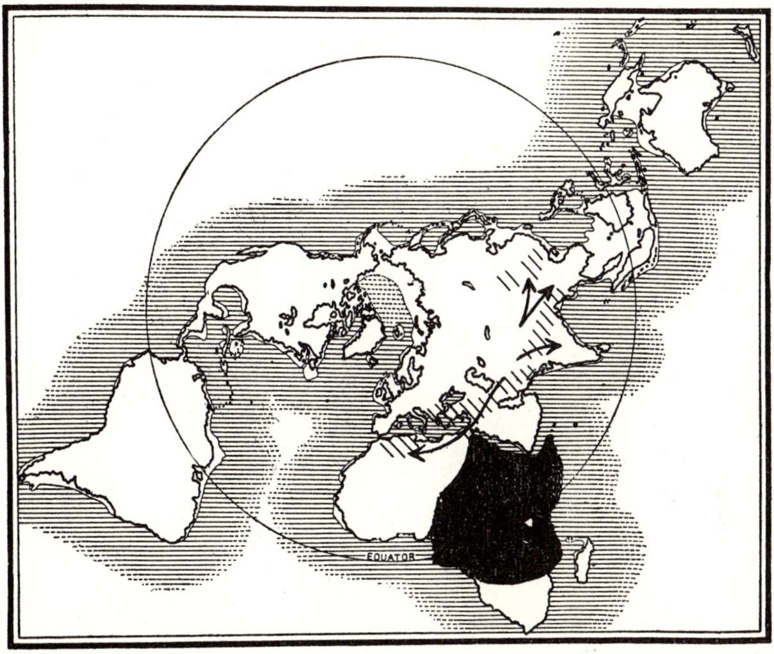

FIGURE 25. Distribution of the Giraffes, existing (solid black) and extinct (shaded)

On present evidence their dispersal center would appear to have been in south central Asia. But the affinities of the Tertiary Giraffidæ to other contemporary ruminants need careful and judicial reconsideration.

The giraffes have approximately the same center of dispersal as the antelopes. This inference from their modern distribution conforms with the geological record. They appear suddenly in the upper Miocene of Europe, but an ancestral series is found in India as far back as the upper Oligocene.[63]

63. See G. E. PILGRIM: Rec. Geol. Sur. India 43: 301. 1913.

TABLE 12

DISTRIBUTION OF BOVIDÆ AND ANTILOCAPRIDÆ

	Neotropical	Nearctic	Palæarctic	Ethiopian	Oriental
Recent	None	*Ovis* *Oreamnus* *Bison* *Antilocapra*	Sheep and Goats ! ! Cattle Antelopes	Cattle Antelopes ! !	Cattle ! ! Antelopes !
Pleistocene	None	*Bison* *Antilocapra* ((*Capromeryx*))	Sheep and Goats Cattle ! Antelopes !	Cattle Antelopes	Cattle Antelopes
Pliocene	None	[*Merycodus*][64]	Cattle Antelopes ! !	(No record)	Cattle ! ! Antelopes ! !
Miocene	None	[*Merycodus*][64]	Antelopes ! ! (late Miocene)		Cattle ! ! Antelopes ! !
Oligocene	None	None	[Ancestral Primitive Ruminants]	None	None

TABLE 13

DISTRIBUTION OF GIRAFFIDÆ

	Nearctic	Palæarctic	Ethiopian	Oriental
Recent	None	None	*Giraffa* *Ocapia*	
Pleistocene	None	None		
Pliocene	? None	None (unless in China)	(No record)	*Sivatherium* *Hydaspitherium* *Giraffa* etc.
Miocene	? None	*Helladotherium* *Samotherium* etc.	(No record)	? *Giraffa* *Progiraffa* etc.
Oligocene	[*Syndyoceras* and *Protoceras*][65]	Ancestral Primitive Ruminants	None	*Progiraffa* (? ancestral to the Giraffidæ)

64. *Merycodus* is a distant relative, combining characters of Bovidæ and Cervidæ.

65. Remote and archaic collateral relatives, family Protoceratidæ. It is by no means certain that *Dromomeryx* and other undescribed genera from the North American Miocene provisionally referred to the Cervidæ and Brachyodont Bovidæ are not related to the Giraffidæ; but on present evidence the dispersal center of the family appears to be India, and their range confined to Palæogæa.

Besides these surviving groups of ruminants, there are several groups which have not survived. The anthracotheres are one of the earliest of these specialized races; I have elsewhere[66] detailed the data upon which may be predicated a North Asiatic center of dispersal for this group. The living hippopotami show a modicum of resemblances to this type, which

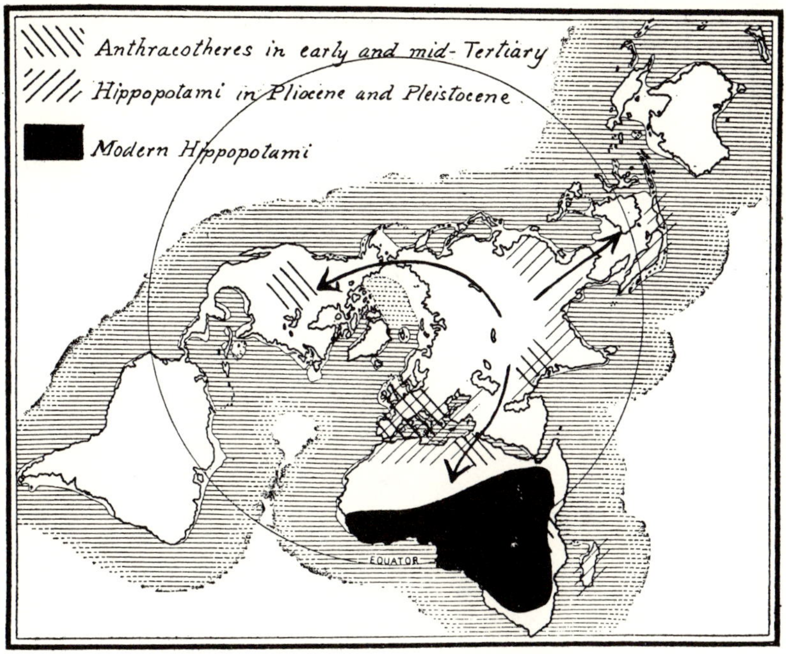

FIGURE 26. Distribution of the Anthracotheres and Hippopotami

The Anthracotheres were a large and widely dispersed group in the Oligocene and Miocene, especially in the Old World, but found also in the Oligocene of North America. The Hippopotami appear to be specialized survivors from the same stock; they are confined to the Old World and their range has been greatly restricted since Pliocene and Pleistocene.

may mean that they are derived from some early members of it. Their present habitat is Ethiopian; but in the Pliocene and Pleistocene their range was far to the northward—even as far as England on one hand and northern India on the other. While the present distribution of the large hippopotamus is Central Africa, smaller and more primitive precursors have been stranded on the one side in West Africa, on the other (now extinct) in Madagascar and also found refuge in the Mediterranean islands

66. Bull. A. M. N. H. **26**: 1-7. 1909.

until the Pleistocene. (The aquatic habits of the hippopotamus have enabled it to reach these island retreats more easily than terrestrial competitors.)

TABLE 14

DISTRIBUTION OF ANTHRACOTHERES AND HIPPOPOTAMI

	Nearctic	Palæarctic	Ethiopian	Oriental	Malagasy
Recent	None	None	*Hippopotamus* *Chœropsis*	None	None
Pleistocene	None	*Hippopotamus* *Chœropsis*[67] (Cyprus)	*Hippopotamus*		*Hippopotamus* (dwarf species)
Pliocene	None	*Hippopotamus*	(No record)	*Hippopotamus* *Hexaprotodon* *Merycopotamus*	(No record)
Miocene	*Arretotherium*	*Brachyodus*	Anthracotheres	*Anthracotherium* *Hemimeryx* *Sivameryx* etc.	
Oligocene	*Anthracotherium* *Ancodus*	*Anthracotherium* *Ancodus* "*Brachyodus*"	*Ancodus*	*Merycops* *Hyoboops* "*Brachyodus*" *Anthracotherium*	(No record)
Eocene	None	*Ancodus*	((No record))		

The remaining groups of ruminants are not of especial interest in this discussion. The entelodonts are Holarctic; the oreodonts Nearctic; anoplotheres and cænotheres Palæarctic; there is no evidence that they originated elsewhere or that they reached any other zoölogical region. *Entelodon* (*sensu lato*) appears simultaneously in Europe and the United States in the beginning of the Oligocene, without direct ancestry in either continent, and is regarded by Peterson[68] as probably from an Asiatic source.

---

67. *H. minutus* is (*vide* Bate) congeneric with the Liberian species. The rules of priority call for the application of *Hyopotamus* Kaup to this genus, instead of *Chœropsis*.

68. O. A. PETERSON: Mem. Carn. Mus. **4**: 145-148. 1909.

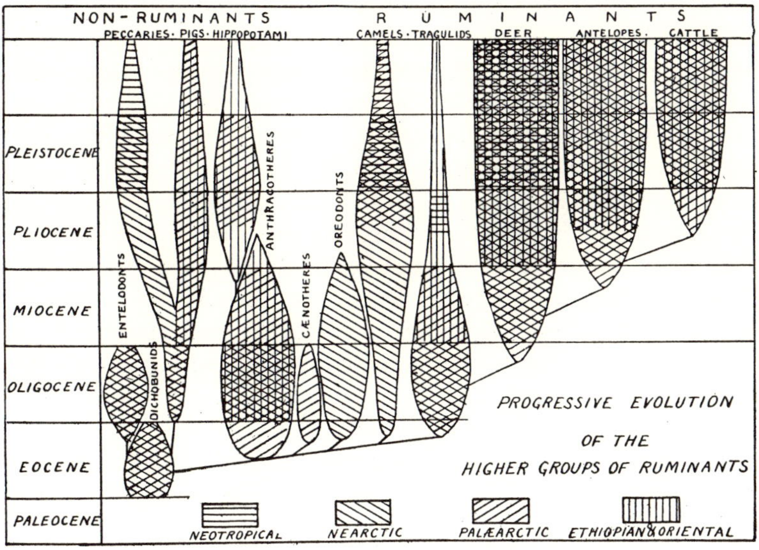

FIGURE 27. Phylogeny and distribution of the Artiodactyla

Most of the families appear to have originated in the Nearctic or Palæarctic region and spread thence outwardly to the more peripheral regions. The higher types are of more recent origin and are still dominant in the Holarctica.

Proboscidea

The later Tertiary and Quaternary history of the mastodons and elephants agrees with the various groups that we have been considering in indicating Asia as the center of distribution of the race. Elephants are now limited to the Ethiopian and Oriental regions, but in the Pleistocene their range was over the whole of Europe, Asia and North America, as well as Africa. The northern species, although of smaller size, are more progressive than the southern species in the specialization of the teeth, proportionate length of tusks, shortening of skull with concomitant elongation of trunk. The more primitive mastodons first appear in India in the Oligocene, in Europe in the lower Miocene, in North America in the middle Miocene. The intermediate stages leading to the mammoths and elephants are best shown in the Pliocene and Pleistocene of India; a less exact series may be found in North America. The mastodons reached South America in the Pleistocene; the mammoths and elephants never reached that continent.[68a] The earlier stages in the phylogeny of the Pro-

[68a. *Parelephas* reached the northern coast of South America. See Henry Fairfield Osborn, Amer. Mus. Novitates **393**: 1929.—E.H.C.]

boscidea have not, however, been found either in Europe or North America but have been recognized in the Oligocene of Egypt. From this fact, it has been generally concluded that the Proboscidea first evolved in the Ethiopian region. But it should be remembered that northern Egypt is not strictly within the Ethiopian region but belongs with all of northern Africa to the Mediterranean subregion of Holarctica. Owing to its proximity to the Ethiopian region, it contains Ethiopian elements in its modern fauna and may have contained more in the past. But it is not clear that the Oligocene Proboscidea must be numbered among these. There is no evidence that their center of dispersal was not Asiatic in early as in later Tertiary; but it must have been too far to the south to admit of their reaching Europe or North America, until after their spread into northeast Africa. We must therefore conclude, apparently, that the dispersal center was transferred to the north and east during the course of the Tertiary—a quite exceptional feature, beside which the question of its original location, whether in southern Asia or in Africa, appears much less important.

TABLE 15

DISTRIBUTION OF THE PROBOSCIDEA

	Neotropical	Nearctic	Palæarctic	Ethiopian	Oriental
Recent	None	None	None	*Loxodon*	*Elephas*
Pleistocene	*Dibelodon*	*Elephas* *Mastodon*	*Elephas* *Mastodon*	*Elephas* ? *Trilophodon*	?
Pliocene	None	((*Stegomastodon*)) *Dibelodon* ((*Rhynchotherium*))	*Elephas* *Mastodon* *Tetralophodon*	(No record)	*Stegodon*
Miocene	None	*Trilophodon*	*Trilophodon* *Dinotherium*	*Dinotherium*	*Tetralophodon* *Trilophodon* *Dinotherium* ((*Hemimastodon*))
Oligocene	None	None		*Palæomastodon* *Mœritherium*	
Eocene	None	None	None	*Mœritherium*	(No record)

Sirenia

The most primitive sirenians are found in the late Eocene of Egypt. As these were apparently contemporary with more progressive types in the Middle and Upper Eocene and Oligocene of Europe, they indicate, if anything, that the Mediterranean shores held a less progressive fauna than the North Atlantic. The Oligocene and Miocene types are approximately ancestral to both the modern groups, manatees and dugongs. Apparently the manatees became characteristic of the North Atlantic, the dugongs of the Indian Ocean shores. The progressive cold of the later Tertiary and Pleistocene has driven the manatees out of the Arctic and northerly Atlantic shores and their northern limit is now Florida on the western, and the African coast on the eastern side. They have not been found fossil north of 40° N. lat. on the American coast,[69] for the excellent reason that there are practically no Tertiary littoral deposits north of that latitude.

The occurrence of *Manatus* in West Africa and in the West Indian and South American coasts is among the arguments used in support of a transatlantic bridge; but there is no evidence at all that the ancestors of *Manatus* did not inhabit the whole of the North Atlantic and Arctic basin during the Tertiary. It is certain that they did inhabit parts of the intervening European and American littoral, and the negative evidence elsewhere is obviously worthless, because there are no formations known in which they might be found.

Condylarthra and Specialized Successors

We may here consider the distribution of a number of extinct groups of Tertiary ungulates or semi-ungulates, whose rise and culmination took place at an earlier epoch and under different conditions from those which we have discussed. The Condylarthra are an extremely primitive group of hoofed mammals, fulfilling nearly the theoretical requirements for the common ancestral type of all placental ungulates. The earliest known artiodactyls and perissodactyls are, however, too much specialized to be immediately derived from the known Condylarthra. Condylarths first appear in the Paleocene of North America and Europe and in South America in the *Notostylops* fauna, here regarded as Eocene. In North America, they develop through the Taligrada into the Amblypoda, culminating in the highly specialized Dinocerata. In South America, they ap-

69. For distribution of manatees during Tertiary *vide* Hay, Bibl. Foss. Vert. N. A., U. S. G. S. Bull. **179**: 583-4, 1902; of Old World Sirenians, Abel, 1904. Abh. Geol. Reichsanst **19**: 214. 1906, Neues Jahrb. **2**: 50-60. 1912, Palæontographica **59**: 292. [Also G. G. Simpson, Bull. Amer. Mus. Nat. Hist. **59**(8): 1932.—E.H.C.]

parently develop during the Tertiary in absence of Artiodactyla and Perissodactyla into a great variety of hoofed mammals, the Toxodontia and Typotheria, Litopterna, Astrapotheria, Pyrotheria. The Arsinoitheria of the Oligocene of Africa, perhaps also the Hyracoidea and Proboscidea, may also be regarded as evolved from primitive Condylarthra, in absence of the higher ungulates of the Asiatic center of dispersal. We have therefore direct or inferential evidence that at the beginning of the Eocene the Condylarthra inhabited the Palæarctic, Nearctic, Neotropical and Ethiopian regions. There is no reason to suppose that they were absent from the Oriental region, but they evidently did not reach Australia or Madagascar.

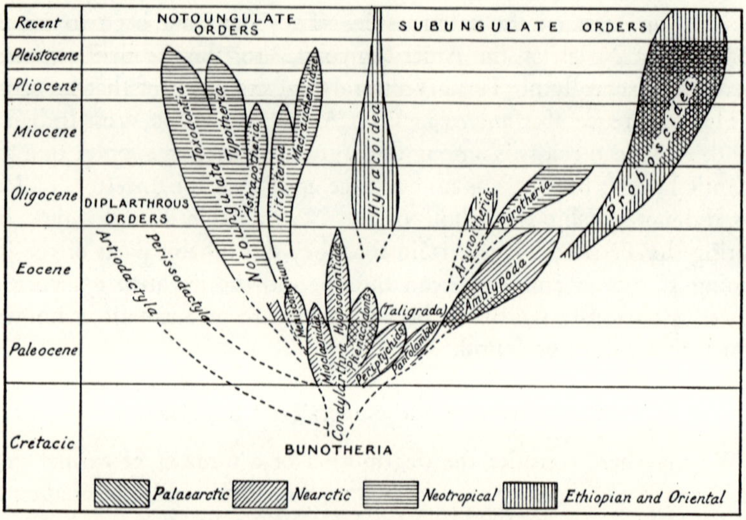

FIGURE 28. **Relationship of the Condylarthra to the Notoungulate and Subungulate groups of hoofed mammals**

In indicating the distribution, Egypt, Syria, etc., have been included with Ethiopia, as the essential facts in this case could thus best be represented. "Bunotheria" are the common ancestral stock (hypothetical) of the Creodonta-Carnivora-Condylarthra-Amblypoda group.

The worldwide dispersal of the condylarths at the opening of the Tertiary (partly hypothetical and exclusive of Australasia and Madagascar) may be regarded as due to the epoch of elevation and disturbance which closed the Cretaceous. The subsequent development of peculiar and highly specialized ungulates during the Eocene in the several great continents is attributable partly to the isolation of these continents during that period due to submergence of the low lying connecting regions, partly to the prevalence of more uniform climatic conditions all over the world and the consequent lack of environmental pressure tending to force a change in habitat. Towards the end of the Eocene began a period of progressively intensified elevation and disturbance, with refrigeration of climate beginning at the poles; this culminated in the Glacial epoch. The northern fauna successively invaded the tropical and southern continents and swept before it nearly all their autochthonic faunæ.

In Africa, we see this invasion in progress in the Oligocene; the anthracotheres, forerunners of the great ruminant invasion have already appeared; to these may yet have to be added *Palæomastodon* as a forerunner of proboscidean invaders (although on the present record the Proboscidea may appear an autochthonic group); while the hyracoids, with *Mœritherium, Arsinoitherium, Barytherium* and some less known types are apparently autochthonic since Paleocene. Unfortunately, our view stops here; we know little of the progress of this invasion until the late Pliocene, when these invaders had themselves disappeared before a succession of later invasions or become modified into new types.

In South America, the isolation lasted much longer, and owing to the great southward extension of the continent, a highly progressive independent center of dispersal was set up in Argentina. Whatever criticisms may be made of the phyletic theories of Dr. Ameghino, so far as they affect the evolution of the mammalian races of the northern world, I think that there can be no question that he has brought out a remarkably complete series of phyla in the autochthonic races of South America. The closeness of these series, and the large amount of progressive evolution which they involve, on lines analogous to those of the northern mammals, are fair indices that the controlling forces were similar and that the southern end of the continent was the chief center of dispersal. The various types of structure which were developed in northern mammals during the Tertiary, in adaptation to the progressive change of environment, are almost all paralleled, occasionally exceeded in degree by these southern races; but they are very generally seen in different combinations, as Professor Gaudry has so clearly shown.[70]

Had the Condylarthra reached Australia, we should expect to find there a group of placental ungulate orders peculiar to the region, like those of Tertiary South America, persisting to the present day. But we find, instead, that the marsupials evolved into the herbivorous fauna. In Madagascar the lemurs may be regarded as filling the place which primitive ungulates would have taken, if they had reached the island; but the case is not so clear.

70. Albert Gaudry: Annales de Paléont. 3: 41-60. 1908.

Edentata

The edentate orders afford among the unguiculates a broad parallel in their distribution and history to the Condylarthra and their successors among the ungulates. Their extinction has been somewhat less complete; a few highly specialized survivors remain in the Neotropical, Ethiopian and Oriental regions.

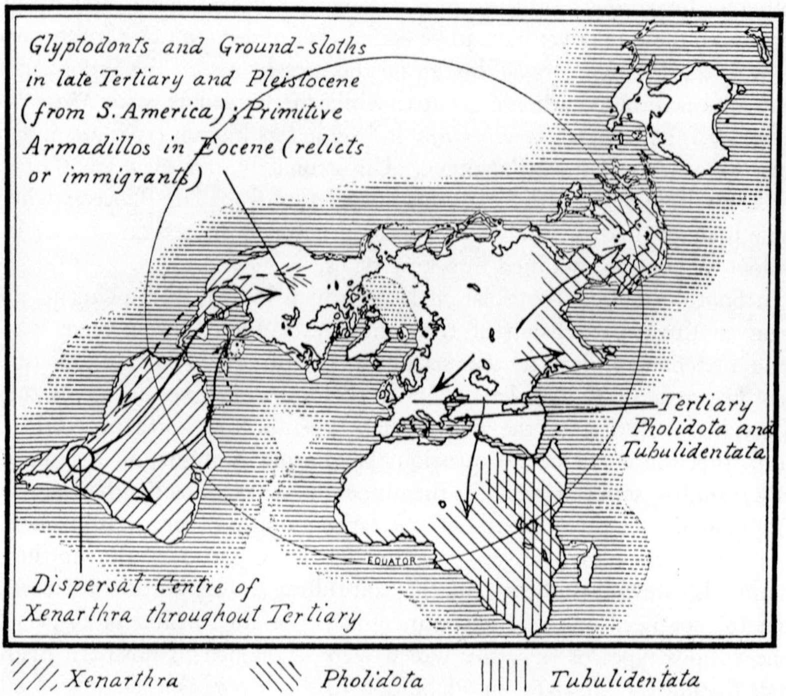

FIGURE 29. Distribution of the edentate orders

The New World edentates or Xenarthra may have originated in Cretaceous North America, but their Tertiary dispersal centers were South American, apparently in or near to Patagonia. The dispersal centers of the Pholidota and Tubulidentata would appear to have been Palæarctic, but very little is known of their fossil record.

The super-order Edentata is an artificial assemblage including the three surviving orders Xenarthra, Pholidota and Tubulidenta and the extinct order Tæniodonta (= Ganodonta). The ((Palæanodonta)) of the Eocene of North America may perhaps be regarded in a broad way as representing the primary type of the Xenarthra ((and Pholidota, but not of the Tubulidentata, while the Tæniodonta are an archaic early specialized side branch which disappeared at the end of the Eocene)). They are far more primitive and nearer to the generalized eutherian type; but they show certain unique Xenarthran peculiarities in foot-construction and in the pelvis, and the dentition in the two known phyla progressively evolves on lines leading towards, although not into, the specialized edentate types.

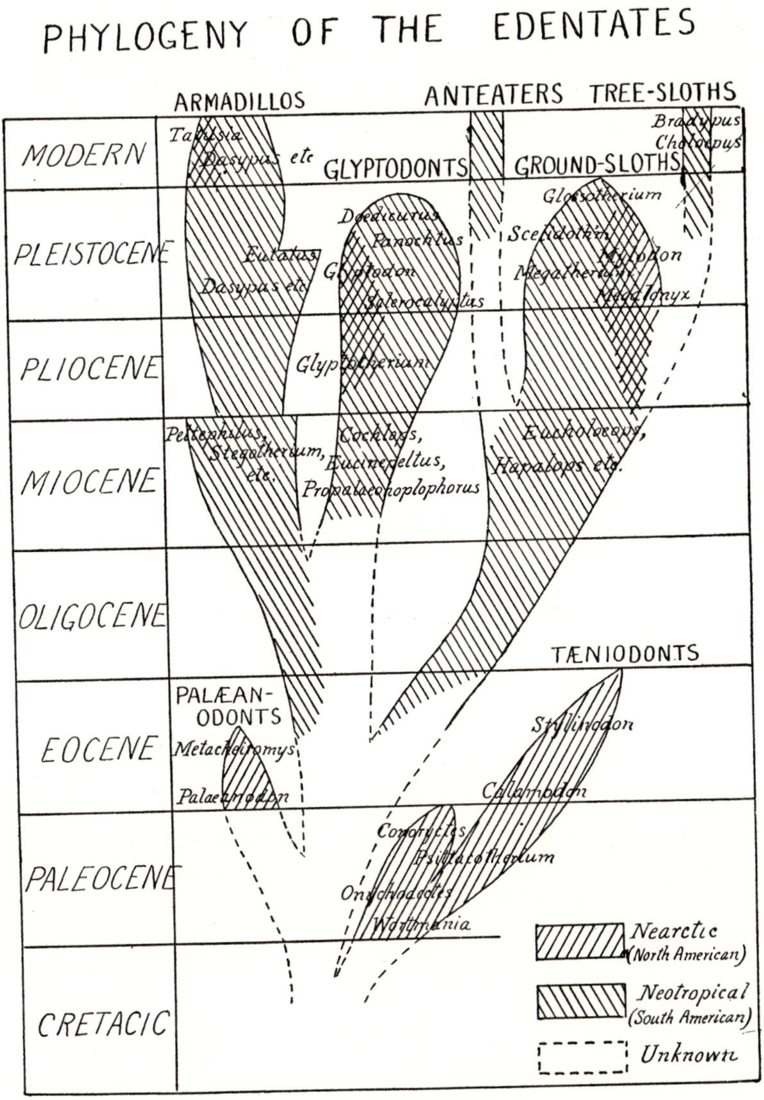

FIGURE 30. Distribution and phylogeny of Xenarthra and Tæniodonta

The aberrant North American groups appear to be relicts indicating a northern origin of the Xenarthra, but the evidence is not conclusive.

The Tæniodonta range from Paleocene to Upper Eocene in North America and are doubtfully recorded in the early Eocene of Europe. They may be hypothetically regarded as a Cretaceous-Eocene ancestral group in the northern world, from whose early members budded off the ancestral Xenarthra in the Nearctic, also the Pholidota in the Palæarctic, the whole group being driven southward at the beginning of the Tertiary, except for a few lingering remnants, rare and little known. ((The Tubulidentata also would seem to represent an archaic disappearing group, but derived from the Condylarthra while the Xenarthra and Pholidota appear to be derived from Pre-Tertiary Insectivora.)) Of these lingerers, we may instance in the (Bridger) mid-Eocene of Wyoming *Metacheiromys,* whose affinities are distinctly armadilloid and ((*Palæanodon*)) in the Lower Eocene of Wyoming approximately ancestral to it ((as well as to the Pholidota, and serving to connect these specialized types with the early insectivores)); *"Lutra" franconica* of the Oligocene of Germany, shown by Schlosser to be related to the Aardvark, *Palæomanis* and *Orycteropus* of the Miocene of Samos, and more doubtfully *Palæorycteropus* and *Necrodasypus* (in part) of the Oligocene of France.

((The)) rare ground-sloth remains from the Lower Pliocene of the western United States are to be regarded ((however not)) as surviving Northern edentates ((but)) as immigrants from the south.

The Old World edentate((s)), although still surviving in Ethiopia and the East Indies (*Manis*), are not known to have undergone any considerable expansion during the isolation-period of the early Tertiary.[71] The Xenarthra, on the other hand, are first represented in the early Tertiary of South America by armadilloid forms and they blossomed out in the isolated continental conditions that prevailed during the Tertiary in that continent into a wide range and diversity of type, just as the Condylarthra appear to have done under the same conditions there and the marsupials in Australia. Of the five principal groups—tree-sloths, ground-sloths, anteaters, armadillos and glyptodonts—only the second, fourth and fifth are known as fossils, and only the first, third and fourth have survived. The fossil groups reached their maximum of size and specialization in the Pleistocene, and reinvaded North America in the Pliocene and Pleistocene (possibly earlier[72]), but only the armadillos have maintained any foothold

---

71. But this may be due only to the imperfection of the geologic record. We know ((little)) of the early Tertiary faunæ of the Ethiopian and Oriental regions, save for the Oligocene of Egypt. The Eocene faunæ of South Africa, India and the East Indies may have included a considerable expansion of pholidate or tubulidentate mammals, corresponding to the xenarthral expansion of the New World, but earlier extinguished because of the earlier invasion of those regions from the north.

72. There is ((much)) question as to the true horizon of the ground-sloth claw found by Sinclair in the Mascall formation (Middle Miocene) of Oregon. The specimen may have washed down from the overlying Rattlesnake Beds, Lower Pliocene [oral communication from J. C. Merriam].

in the northern world until modern times and these only in the southwest corner of the Sonoran region. The anteaters and tree-sloths might be expected to have originated in Patagonia and to have been driven northward to tropical South America in accord with the theory of climate and evolution here advocated. The geological record, however, has failed to show any certain evidence of this, and, as the Patagonian record is a comparatively full one, this fact should be counted as evidence that climatic change is not the only causal factor of evolution. We must suppose, if the record be adequate, that these groups originated and evolved in tropical South America. The armadillos are an extremely persistent group, and the record gives no really convincing evidence of a Patagonian dispersal center, although it might be so interpreted.

Glyptodonts and ground-sloths appear in the Pliocene and Pleistocene of North America. The Pleistocene genera except *Megalonyx* are closely allied to the genera of the Pampean formation, in part identical therewith (*Brachyostracon, ? Glyptodon, Chlamydotherium, Megatherium, Megalonyx, Nothrotherium, Mylodon*). These, or allied genera equivalent in specialization, inhabited South America from Ecuador to Patagonia in the late Pliocene and Pleistocene. The only genera found in the Pliocene of North America are *Megalonyx* and *Glyptotherium,* decidedly more primitive and are best interpreted as earlier forerunners of the main invasion which appeared at the beginning of the Pleistocene. *Mylodon* has been recorded from the Blanco beds of Texas, but this is an error.

Marsupialia

Marsupials are at present almost limited to the Australian and Austro-Malayan region, where, in the absence of placental mammals, they have diversified into a wide variety of size, habits and adaptation, paralleling the adaptive radiation of the higher mammals in the northern continents. A single unspecialized group, the opossums, representing quite nearly the primitive type from which all marsupials are derivable, survives in the Neotropical region, one or two of its species ranging northward into the Sonoran subregion of Holarctica. Another primitive survivor in the Neotropical region is the rare little *Cænolestes.* ((Osgood, who has made a most careful study of this genus, regards it)) as a primitive member of the diprotodont marsupials. ((It has been)) considered ((by others)) to be of polyprotodont affinities, its diprotodont resemblances being due to parallelism.

What we know of the paleontology of the order is in complete accord with the theory of their being primarily of northern origin, their dispersal preceding that of the early placentals.

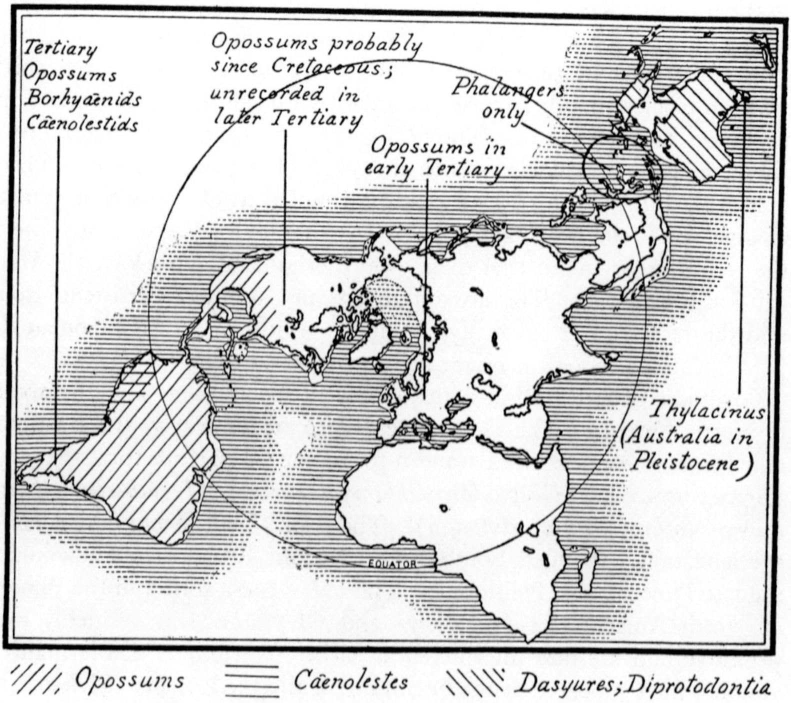

Figure 31. Distribution of Marsupials

This is probably to be regarded as due to a very ancient dispersal from the north, followed by differentiation and dispersal during the Tertiary of specialized adaptations parallel in the Neotropical (Borhyænids and Cænolestids) and Australian regions (Thylacine-Dasyures and Diprotodonta). The Phalangers of the Austro-Malayan islands are regarded as marginal types from an Australian dispersal center.

The fragmentary and little known mammals from the Mesozoic formations of Europe and North America were in large part marsupials, so far as we can judge from what is known of them.

The most distinctive group among them were Multituberculata or Allotheria. Gidley[73] ((in)) (1909) brought forward strong evidence for the view that these animals were an archaic, early specialized branch of the marsupials paralleling the later diprotodonts. ((Later discoveries and researches)) indicate that the relationship was not as close as had appeared. Dr. Broom has even maintained that these animals were nearer to monotremes than to marsupials, but in my judgment he has failed to adduce any really valid evidence for this view. But while they are in the Metatherian stage of evolution I do not think they can be included in the order

73. J. W. Gidley: Proc. U. S. Nat. Mus. 36: 611-626. 1909.

Marsupialia on the data now available.[74] They occur (doubtfully) in the Rhætic of Germany, certainly in the Upper Jurassic and Lower Cretaceous of England (*Plagiaulax*[75]) and Wyoming (*Ctenacodon*). They again appear in the uppermost Cretaceous and Paleocene of North America (Lance formation of Wyoming, Fort Union of Montana, Puerco and Torrejon of New Mexico) and Europe (Cernaysian) in the genera *Ptilodus, Neoplagiaulax, Polymastodon* and *Meniscoëssus.* They are questionably recorded in the Eocene *Notostylops* beds of Patagonia, in the genera *Propolymastodon, Polydolops,* etc. (which more probably belong to the same group as *Cænolestes*). They are not known elsewhere except for part of a jaw from the middle Cretaceous (Belly River) of Canada and a jaw (*Karroomys*) from the Jurassic (Karroo beds) of South Africa. The front half of a skull long ago found in the Karroo beds and described as *Tritylodon* is probably to be referred to this group, although its mammalian nature has been questioned.[76] ((*Stereognathus* of the middle Jurassic of England, appears to be related to *Tritylodon*.[77]))

In addition to the Multituberculata, there are in the Jurassic and basal Cretaceous of England and Wyoming a number of mammals with simpler and more numerous teeth whose affinities are very uncertain. Whether they are ancestral to marsupials, to placentals, to both or to neither is, in the writer's opinion, an unsettled question. Its definite solution must probably await the discovery of more complete material.

In the uppermost Cretaceous (Lance formation) of Wyoming are found in addition to teeth and jaw fragments of Multituberculata, a variety of tritubercular teeth, some associated with fragments of the jaw. These appear to be more definitely referable to the polyprotodont marsupials; some of them may be quite near to the opossum. ((*Eodelphis* of the Belly River, is the best known of these Cretaceous marsupials, and it is with difficulty separable from the Didelphidæ.)) I have seen no evidence among them of placental mammals, although most of them are too fragmentary to exclude the possibility of the presence of Eutheria.

The Paleocene fauna of New Mexico, Montana and France contains numerous placentals, a few Multituberculata, ((and opossums;)) but no

---

[74. See Walter Granger: Bull. Geol. Soc. Am. **26**: 152 (abstract). 1915. Walter Granger and George Gaylord Simpson, Bull. Amer. Mus. Nat. Hist. **56**: 601-676. 1929.—E.H.C.]

75. *Bolodon* is a synonym of *Plagiaulax,* fide Gidley.

76. Broom has recently made a careful restudy of the affinities of *Tritylodon,* and concludes that it is a mammal, but not closely related to the marsupials, and represents an archaic specialization with many primitive characters inherited from the cynodont reptiles. R. Broom: Trans. S. Af. Phil. Soc. **17**: 73-77. 1905. Proc. Zoöl. Soc. London 760-768. 1910. Bull. Am. Mus. Nat. Hist. **33**: 115-134. 1914.

((77. The jaw on which *Stereognathus* is founded, is an upper jaw, although it has been described as a lower jaw by Owen, Osborn, Marsh, and even recently by Petronievics.))

true diprodont marsupials have yet been positively recognized in it. It is evidently not derived (except for certain of the Multituberculata ((and presumably the didelphids))) from the fauna of the Lance formation. Yet it is almost, perhaps quite, contemporaneous with it and must be supposed to represent a distinct facies of the fauna, differing in habitat from that of the Lance formation (the Fort Union is partly intermediate). Polyprotodont marsupials certainly persisted in North America and Europe, for we find the remains of species nearly related to the existing opossums in the Lower and Middle Eocene of Wyoming, in the Oligocene of Colorado ((and Oregon)) and in the Upper Eocene to lower Miocene of France and Germany. They are not known from any later formation in any of the Northern continents.

In the Southern continents, they assumed a much more important position. In South America, in the absence of placental Carnivora, the polyprotodont marsupials developed into a number of large and small predaceous mammals (Borhyænidæ), so closely paralleling some of the predaceous marsupials of Australia that they have been referred to the same family (Thylacinidæ). Pseudo-diprotodont marsupials were also fairly common, taking the place in the fauna held by Insectivora in the North, this group of placentals (except for a single type) not having reached South America. The marsupials of South America did not develop into groups taking the place of northern ungulates, rodents or primates, since primitive placentals of these groups (Condylarthra, ? Hystricomorpha, ? Lemuroidea) had penetrated into South America before it was separated from the Northern world, and there developed along lines sub-parallel to the development of the higher placental groups in the North, but distinct and less progressive.

In Australia, the marsupials assumed a still more important position, as the only mammals of that continent. The placental mammals of the northern Tertiary did not reach Australia, except for a few strays—bats and mice and the dingo—which were too few in numbers and of too recent introduction to affect seriously the course of mammalian evolution on that continent. In the absence of placentals, the marsupials developed into a wide variety in size, form and habits of life, partially paralleling the higher mammals.

The near resemblance between the modern Australian *Thylacinus* and the Borhyænidæ of Tertiary South America has been used as an argument for an Antarctic connection between the two. Such a hypothesis will not bear close examination. The resemblance is not closer than between parallel adaptations in distinct families of true Carnivora, whose genealogy has been more or less completely traced back through independent lines of descent from unspecialized common ancestors. It is not closer, for instance, than that between the Oligocene Felidæ and the modern *Cryptoprocta* of Madagascar, whose common descent from an unspecialized placental carnivore (viverrid or miacid), analogous to the marsupial didelphyids, is generally admitted. The common characters distinguishing thylacinids and borhyænids from the didelphyids are, without exception, such as would naturally be assumed independently in adaptation to predaceous terrestrial life and have been so assumed in numerous independent parallel adaptations of the same sort among placental Carnivora. On the other hand, *Thylacinus* has retained certain didelphyid characters which are already lost by the most primitive of the Borhyænidæ (palatal vacuities, posterior position of the orbits,[78] an external lachrymal duct, double perforation of the basisphenoid), while in other features (brain development, cursorial specialization, etc.) it is more progressive. The Borhyænidæ are more progressive in the reduction of the last molar, in the differentiation of enamel from dentine, less so in the cursorial adaptation of the limbs and feet.[78a]

Descent from a common ancestral type is undoubtedly shown, but some at least of the above differences point back to Didelphyidæ as this common type. The characters which Sinclair uses to separate the thylacines are the reduced number of incisors, the carnassial specialization of the molars and especially the loss of the metaconid. Every one of these features, besides numerous other common characters which he does not specify, may be paralleled in two or more distinct lines of Carnivora whose common ancestors are not more predaceously specialized than *Didelphys*. The loss of the metaconid occurs in *Cyon, Ischyrocyon, Simocyon* and *Enhydrocyon* among the Canidæ, in all the post-Oligocene Felidæ, in *Gulo, Megalictis, Mustela,* etc., among the Mustelidæ, in the later Hyænidæ, in *Hyænodon* and *Pterodon* among the Hyænodontidæ, in *Patriofelis* among the Oxyænidæ, in all the later Mesonychidæ. Each one of these genera is independently descended from genera in which the metaconid is well developed. In every case, it is simply a stage in predaceous adaptation of the molars, nor can it be assigned any other significance in the marsupial carnivores. There is, in short, no evidence for assuming a closer affinity between thylacines and borhyænids than common descent from didelphyid ancestors, and there is strong evidence against such an assumption. But if this be true, these animals afford no evidence for Antarctic connections between the southern continents; for we have seen

78. Interpreted by Sinclair as a progressive character in *Thylacinus,* but certainly the reverse in analogous placental adaptations.

[78a. See H. E. Wood, Bull. Amer. Mus. Nat. Hist. **51**(4): 1924. Also W. D. Matthew, Proc. Zoöl. Soc. London **1927**: 983. 1927.—E.H.C.]

that didelphyid marsupials were certainly present in the Mesozoic and early Tertiary of Holarctica and of South America, and we have no reason to believe that they would have had greater difficulty in reaching Australia in the Mesozoic or early Tertiary than the murine rodents found at a later date.

The supposed presence of diprotodont marsupials in the South American Tertiary and in modern Australia has also been used in support of Antarctic connections between the two continents. The recent morphologic studies of Dederer[79] and Broom[80] have shown that *Cænolestes* is not a true diprotodont, but in fact belongs to the polyprotodont division of the Marsupialia,[((80a))] and with this genus must be associated all of the Epanorthids and probably all of the so-called Paucituberculata of the South American Tertiaries. If then the Diprotodonta, so dominant and so widely varied in Australia, were wholly absent from South America, while parallel adaptations were developed there from the Polyprotodonta, the distribution of these marsupials affords a valid argument against instead of for any Antarctic connection during the Tertiary.

In view of the great amount of adaptive divergence seen in the various Pleistocene and modern genera of Australian Diprotodonta, the origin of the suborder in Australasia or its earliest invasion of that zoölogical region, must be dated far back in the Tertiary. On our present evidence it may well be regarded as wholly autochthonic, derived from early Tertiary or possibly from late Mesozoic polyprotodonts. Nevertheless, in view of the defectiveness of the Mesozoic record, where we should chiefly expect to find this group, if anywhere in the North, and the presumable rarity of Tertiary survivors, there is nothing unlikely in the view that they originated primarily in the North like their polyprotodont and allotherian relatives and were driven southward with the former group and somewhat more thoroughly extinguished in the North, while in Australia they blossomed out into a great adaptive expansion paralleling the absent ungulate mammals.

It is probable that the opossums survived in North America throughout the Tertiary, although there is no clear record of them in our Miocene and Pliocene.[81] But we know only a small part of our Pliocene fauna as yet, and the Miocene, although better known, represents chiefly the ani-

---

79. Pauline H. Dederer: Amer. Nat. **43**: 614. 1909.

80. R. Broom: Proc. Linn. Soc. N. S. W. **36**: 315. 1911.

((80a. But see W. H. Osgood: Field Mus. Publ. Zool. XIV, 1: 1-15. Jeanette B. Obenchain: *ibid.* 3: 175-232. 1925. E. Lönnberg: Arkiv för Zoologi 14. 1925.))

81. A very badly preserved skull from the Colorado Miocene and a jaw fragment from the South Dakota Miocene in the American Museum collections are perhaps marsupials; but I have never been able to see in either specimen satisfactory proof that they were so, and have consequently never recorded them.

mals of the open plains, the forest fauna being very incompletely represented. On the other hand, it seems probable that the apparent disappearance of marsupials from Western Europe after the Lower Miocene was real, and it is probable that they had disappeared even earlier from Asia. They have not been found in the later Tertiaries of India or China, so that they must have been rare if not absent at that time. The Eocene Tertiary of Asia, where they might be expected to be common, is altogether unknown.[82]

TABLE 16

DISTRIBUTION OF POLYPROTODONT MARSUPIALS

	Neotropical	Nearctic	Palæarctic	Ethiopian	Oriental	Australian
Recent	Didelphyidæ *Cænolestes*	*Didelphys*	None	None	None[86]	Thylacinidæ Dasyuridæ
Pleistocene	Didelphyidæ	*Didelphys*	None	None	None	Thylacinidæ Dasyuridæ
Pliocene	Borhyænidæ[83] Didelphyidæ Epanorthidæ	None known	None	(Record inadequate)	None	(No Mesozoic or Tertiary record)
Miocene	Borhyænidæ Didelphyidæ Epanorthidæ	?[85]	None after lower Miocene	(No record)	None	
Oligocene	Borhyænidæ Didelphyidæ	*Peratherium*	*Peratherium*	None	(No Mesozoic or Early Tertiary record)	
Eocene	*Polydolops*, etc. Borhyænidæ Didelphyidæ *Caroloameghinia*	((*Peradectes*))	*Peratherium*	(No Mesozoic or Eocene record)		
Paleocene		((*Peradectes*))				
Cretaceous	*Proteodidelphys*[84]	Cimolestidæ *Thlæodon* ((*Eodelphis*))	(No record)			
Comanche		(No record)				
Jurassic		Triconodontidæ				

82. The earliest Asiatic Tertiary fauna is that of the Bugti beds of India, lower Burdigalian or upper Aquitanian according to Pilgrim, Rec. Geol. Sur. India, vol. xliii, pt. 4, pp. 264-326. It is therefore either late Oligocene or early Miocene.

83. Fragmentary remains, referred to Hyænodontidæ by Dr. Ameghino.

84. Jurassic, fide Ameghino.

85. Doubtful fragments of jaws which may be Didelphyid.

86. Except on borders of Australian region.

TABLE 17

DISTRIBUTION OF DIPROTODONT AND PSEUDODIPROTODONT MARSUPIALS AND ALLOTHERIA (MULTITUBERCULATA)

	Paucituberculata	Allotheria			Diprotodontia vera
	Neotropical	Nearctic	Palæarctic	Ethiopian	Australian
Recent	*Cænolestes*[87]	None	None	None	Macropodidæ Phaseolomyidæ, etc.
Pleistocene		None	None		Diprotodontidæ Macropodidæ Thylacoleonidæ, etc.
Pliocene	Epanorthidæ[88]	None	None		
Miocene	Epanorthidæ[88]	None	None		
Oligocene	Epanorthidæ[88]	None	None	None	*Wynyardia*[89]
Eocene	*Polydolops*, etc.[88]	?	None		
Paleocene		*Ptilodus* *Polymastodon*, etc.	*Neoplagiaulax*		
Cretaceous		*Ptilodus* *Meniscœssus*, etc.			
Comanche					
Jurassic		*Ctenacodon*, etc.	*Plagiaulax*, etc.	*Tritylodon*	
Triassic			*Microlestes*	*Karroomys*	

Monotremata

The monotremes are the lowest group of mammals, far removed structurally from any others. Their connection with the main stock must

87. This genus is a pseudodiprotodont, as its real affinities are with Polyprotodontia, as shown by Dederer and Broom, *l. c.*

88. Affinities probably with *Cænolestes*.

89. Combines polyprotodont and diprotodont characters.

date back to the end of the Paleozoic Era. Nothing is known of their evolutionary history. The Multituberculata of the Mesozoic and Basal Eocene are regarded by Broom as ancestral to them, but this view is not supported by additional evidence since obtained. *Xenotherium*[90] of the North American Oligocene, referred by its describer to the monotremes, is an Insectivore related to the Chrysochloridæ; *Scotæops*[91] of the South American Tertiary is an Armadillo,[92] and other genera referred by Ameghino to the monotremes probably also pertain to other groups. We find them to-day limited to the Australian region, and surviving even there only by virtue of unusual specializations of habit; *Echidna* protected by its coat of spines, *Ornithorhynchus* by its amphibious habitat, both genera burrowing and nocturnal. Presumably, these genera represent the last relic of the early Mesozoic dispersal movements of the Mammalia.

Summary of the Evidence from Dispersal of Land Mammals

The foregoing review of the several groups of land mammals shows that the more recently evolved and dominant races of Mammalia are to-day mainly Holarctic, and many of them have not yet reached the more peripheral regions; that the ancestry of all these dominant races has been found in the Holarctic Tertiary formations, sometimes in Europe, sometimes North America, more generally a series in each country of equivalent approximately ancestral stages. Where the geological record is adequate, these races are shown to be newcomers in the peripheral continents which they have invaded, and any ancestral series is absent. Their representatives in the peripheral continents are to a varying degree primitive and allied to earlier stages in the evolution of the race as represented in the Tertiary record of Holarctica, but they have specialized more or less along parallel or divergent lines from the direct line of descent of the northern representatives.

When the parallel series in Europe and North America are sufficiently complete they are seen to be not parallel phyla of independent local evolution, but periodically recruited by more progressive new stages, apparently from a common center of dispersal. The relations are like those of one side and the other of a branching tree whose trunk region is unknown to us.

---

90. Earl Douglass, 1905. (The name is preoccupied by *Xenotherium* Ameghino, 1904, a genus of typotheres.)

91. Fl. Ameghino, 1887.

92. W. B. Scott: Rep. Prin. Exp. Patag. 5: 12. 1903.

The more ancient and primitive groups of the Mammalia have mostly disappeared, or are in process of disappearance, from Holarctica. In the peripheral continents, they have undergone in many cases a notable local adaptive radiation and expansion, extensive in proportion to the isolation of these continents from the northern realm, more complete during the early and middle Tertiary than now. When the reunion to Holarctica permitted the northern fauna to invade the peripheral continents, these autochthonous groups were in general unable to maintain themselves against the competition of the more progressive northern races, and have either wholly disappeared or left a few scattered survivors, mostly aberrant specializations which did not come directly into competition with the invading races. The survival of the major part of the marsupial radiation in Australia is attributable to its continued isolation. The apparent fact that Neotropical races of Edentata were able to invade North America during the Pliocene and Pleistocene may be ascribed to two factors:

(1) No Nearctic groups of closely analogous specialization existed at that time.

(2) Owing to the far southerly extension of South America, the evolution of mammals in that region was, so far as controlled by climatic change, more progressive and more nearly equivalent to the Holarctic evolution than in Australia or Africa. Its products therefore were better able to maintain themselves against their northern competitors.

If we regard the Proboscidea as of Ethiopian origin, we must suppose that they too constitute an exception to the general rule that the races evolved in the peripheral regions have been unable to invade Holarctica. But the recent discoveries of Pilgrim and Cooper in the Oligocene of India tend strongly to show that the Proboscidea were from the first, as they certainly were in the later Tertiary, a group of Asiatic, not African, dispersal.

The dominant influence of climate in controlling the range of modern mammals has been emphasized by C. H. Merriam. The mammals adapted to north temperate or even boreal climate are the most specialized and last evolved members of their respective races. The most primitive survivors of northern races, and surviving members of races formerly abundant in the north, are met with chiefly in tropical regions. Similar relations are seen in the faunæ of the Antarctic as compared with the southern tropical regions, although less obvious. This is especially seen in South America. It is displayed there quite as clearly in races, such as the cricetine rodents, cervidæ, etc., which are admittedly of Northern origin, as it is in any autochthonous groups. Hence, it cannot be attributed to a general Antarctic dispersal center, but must be explained as a parallel evolution under similar climatic stimulus.

The general distribution of Mammalia on these lines is almost universally accepted; but many writers have pointed out certain supposed exceptions and found it necessary to account for them by various hypothetical continental bridges. A careful consideration of these supposed exceptions shows that, if due allowance be made for parallelism and for the imperfection of the record, each one can be more satisfactorily interpreted in accordance with the general law. And the acceptance of any such continental bridges would entail migrations of other groups which assuredly have not occurred. The hystricomorph rodents of South America afford a single exceptional instance, in which over-sea transportation from Africa appears to be the only reasonable interpretation of the evidence at hand.

I place much greater weight on the evidence from mammalian distribution than on that of any other terrestrial group for several reasons, as follows:

(1) Their past history, the time, place and method of evolution of the various races, is better known than in any other group of land animals or plants.

(2) The complexity of structure in the hard parts which are preserved as fossils is greater, affording a larger amount of evidence by which we may distinguish parallel or analogous races and determine the closeness of their real affinities. As Stehlin[93] has recently observed, a single tooth of a mammal affords as much structural evidence whereby to determine its relationships as the entire skeleton of most invertebrates. Where our evidence is thus limited (to a single tooth, for example), we may, and frequently do, find difficulty in deciding the exact affinities of a fossil mammal. But where we have the skull or the skeleton or even the entire dentition, the results are correspondingly sure and precise as the data are more extensive.

(3) Owing to their nearness to ourselves, their large size and other causes, we are better able to understand their adaptation and observe and appreciate the factors which may affect their evolution and migration.

In dealing with the evidence furnished by the lower vertebrates and invertebrates, we are hampered by the wider limits of time within which the migration may have taken place, by the relative simplicity of the structure of the hard parts, which makes it less easy to distinguish parallelism from immediate affinity,[94] by the relative scarcity of fossils as compared with living species (among land animals), and by our less certain knowledge of the causes which may control their evolution, their means of migration, and their true evolutionary history and affinities.

---

93. "Über die Säugethiere der Schweizerische Bohnerzformation." Verh. Schw. Naturf. Gesell. 93 Jahresvers 1910, Basel. Page 11 of separate.

94. It may be noted in illustration of this point that a natural cast of the entire carcass of a mammal would afford far less secure information as to its real affinities than would a fossil skull, and less even than a lower jaw with reasonably perfect teeth. The parallel adaptations so frequently recognized among mammals lead to superficial resemblance of distantly related types whose true affinities are readily recognized by the internal structure. If, as among most invertebrates, we had only an external skeleton to guide us, the real affinities would not be so securely recognized.

Interpretation of Negative Evidence in Fossil Mammal Faunæ

In considering a Tertiary mammal fauna, we must keep in mind the facts that there may be large facies of it that are represented imperfectly, if at all, in our records, and that there may be important parts of it which have left little or no record, owing to their habitat, small size or other circumstances. We may, with some reserve, conclude that the entire absence from the record of a group which is abundant in other faunæ indicates its real absence from the fauna. But we are not justified in so concluding in the case of rare or inconspicuous races. It is fair to assume that the absence of Perissodactyla from the Oligocene fauna of Egypt or the Miocene fauna of Patagonia was real, and not a matter of defective record. The same assumption would be unjustified in the case of didelphid marsupials and dilambdodont Insectivora respectively. But the most conclusive evidence of the absence of a certain group from a given fauna is that while it is not found fossil, another group is found to have become adapted on parallel lines, taking its place in the fauna. The absence of Perissodactyla and Artiodactyla from the Miocene of South America is confirmed by our finding Litopterna, Toxodontia and Astrapotheria, which parallel in adaptation the horses, rhinoceroses, tapirs, camels, etc., of the North; the absence of Carnivora by the parallel adaptation of marsupials to take their place. The evolution of lemuroid primates in Madagascar into large quadrupedal forms apparently paralleling certain groups of Ungulates,[95] affords some evidence that the Tertiary hoofed mammals were unable to invade Madagascar.

The absence of fissipede Carnivora from the recorded Oligocene fauna of Egypt would not be conclusive in itself; but, coupled with the exceptional variety and abundance of the more archaic creodonts of the family Hyænodontidæ, it is very strong evidence that fissiped Carnivora had not yet invaded the Ethiopian region, at least in any considerable numbers.

DISPERSAL OF REPTILIA

The essential adaptive feature which distinguishes mammals and birds from the reptiles out of which they arose lies in the non-conducting covering to the skin,—of hair or fur among mammals, of feathers among birds. The assumption of this covering enabled the body to be kept at a

95. The skull and the short limbs of *Megaladapis* are very suggestive of such types as *Promerycochœrus*. The feet do not, however, indicate a terrestrial habitat, nor are the teeth efficient in grinding. The resemblance in teeth and skull of *Archæolemur* to the Anthropoidea is very marked.

uniformly high temperature, thus favoring the maximum of bodily activity, and making it practicable to develop the circulation and the entire organization to a much higher standard. It also made these classes of animals independent of the temperature of their environment. It enabled them to withstand cold or variable climate and to take full advantage of the conditions of the colder regions, which appear to favor a higher development than can be attained in moist tropical countries.

The initial development of mammals and birds took place, so far as we are able to judge, during the great arid period of the Permian-Triassic. They appear to have been derived from unknown groups allied respectively to the theromorphous reptiles and to the ornithischian dinosaurs. We know almost nothing of their Mesozoic evolution, because the upland epicontinental formations of the Mesozoic, in which this record should be chiefly preserved, have been totally swept away, or if any remnants remain, they have not been recognized and sufficiently explored to recover it. The formations of the swamps and coastal marshes, river-deltas, littoral regions and shallow seas of the Mesozoic are extensively preserved and their inhabitants well known to us. But of the upland fauna, we get only an occasional glimpse in such deposits as those of Solenhofen, where a few remnants of the fauna of the adjoining uplands have been preserved in great perfection. We have, indeed, indirect evidence as to the nature of the upland fauna of the Mesozoic, for the successive groups of swamp dinosaurs, the marine birds and pterodactyls of the later Mesozoic and the abundant and varied mammalian fauna which appears at the beginning of the Tertiary are not derivable, any of them, from their predecessors in the swamp or marine faunæ, but must be traced back to ancestors distinctly adapted to dry-land life, which reinvaded the coast-swamp, littoral or marine provinces. This will appear more in detail in the discussion of the several orders. The point here to be emphasized is that the dry-land vertebrate fauna has been throughout the dominant facies and has repeatedly reinvaded the swamp and sea-coast provinces, the higher activity and better organization acquired on land giving its members, when readapted to the marsh or littoral conditions, an advantage which enabled them to supersede the autochthonous dwellers in those conditions. *Per contra,* there have not been a succession of invasions of the dry land by the vertebrate inhabitants of swamp and sea-coast. Once established on dry land, the primary groups of dry-land reptiles held their own and evolved and expanded into higher types and greater variety, but they were not recruited, so far as the evidence shows, by new invasions from the swamp and aquatic fauna.

Dinosauria

The dinosaurs appear to be primarily a dry-land adaptation (properly speaking, two distinct but parallel adaptations) of the primitive reptiles.[96] Their most obvious adaptive characters lie in the long limbs and swift-running gait and the general parallelism to the ratite birds. As such, the conditions of life would tend to greater activity and higher development and enable them, when they reinvaded the swamps during the epochs of great swamp-extension, to reach greater size and dominance. It is these readaptations that are chiefly known to us and are apt to give the idea that the dinosaurs were distinguished by gigantic size and massive proportions. In fact, these are no more typical of the order as such than the whale, hippopotamus and elephant are fairly typical of the mammals as such. There must have been multitudes of small dinosaurs, mostly inhabiting the upland, a smaller number living among the swamps and marshes, but we know comparatively little about them. Some notion of their numbers and variety in the Triassic is gained from the innumerable footprints spread over the Triassic shore-deposits of the Connecticut River. But of all this multitude, we have actual remains of only two or three types. The *Compsognathus* skeleton of Solenhofen is, perhaps, an example of the small light-limbed upland dinosaurs of the Jurassic; *Hallopus* and *Podokesaurus* are perhaps fairly representative of their Triassic ancestors. The Jurassic sauropods, while highly specialized for aquatic life and river-bottom wading, yet retain a few features indicative of former land life. One of these is the long limbs, which it would seem must have been acquired on land. Another is the fact that the knee bends forward as it does in all other dinosaurs, while in reptiles primarily amphibious the knee bends outward and the limbs are short. The elbow of the Sauropoda, on the other hand, bends outward, as in reptiles generally, not backward, as it does in primarily quadrupedal land animals, and this I take to be an indication that their quadrupedal gait is partly secondary and that they are derivable from long-limbed, partly bipedal ancestry. The shortening of the feet and pillar-like construction of the limbs is an obvious parallelism with the specialization of these parts seen in all large land mammals and is an adaptation to their great size. No near parallel can be found to this group among living animals; the hippopotamus affords some suggestions, but diverges widely in many respects.[97]

---

96. F. von Huene: Geol. u. Pal. Abh., N. F. **13**: 22-38. 1914; Neues Jahrb., Beil. **37**: 577-587. 1914.

97. See W. D. Matthew: "The Pose of the Sauropodous Dinosaurs," Amer. Nat. **44**: 547-560. 1910.

I have already referred to the primary adaptation of the dinosaurs as a dry-land adaptation of the Reptilia. To a limited extent, the modern lizards represent a corresponding adaptation but not carried so far or occupying so important a place in the fauna. The lizards have to compete with the large and varied dry-land fauna of mammals, and relatively to these, they occupy but an unimportant niche in the terrestrial life. They suggest, however, the sort of animal which in the absence of a higher competing type evolved into the dinosaurs, and their more specialized types (*e. g., Chlamydosaurus*) mimic them in proportions in a most instructive manner.

Dinosaurs are first recorded from the Triassic; those which we actually know[98] are of moderate to large size, slender and long limbed as compared with other reptiles, not highly specialized in dentition, unarmored and some but not all bipedal in gait. Indirect evidence in the multitudes of footprints of the Connecticut Valley sandstones shows that there must have been also a great number and variety of small bipedal three-toed forms all presumably dinosaurs, and other reptiles with shorter feet and more numerous toes which may also have been dinosaurs, although not generally so referred. Lull[99] states in regard to the latter: "These forms seem to represent survivors of the ancient stem from which the dinosaurs arose; they may, however, represent primitive quadrupedal dinosaurs which had not yet acquired the erect gait." He calls attention to their possible relationship to *Protorosaurus* and *Kadaliosaurus*.

---

98. The principal references on Triassic dinosaurs are the following:

R. Broom: "On the South African Dinosaur Hortalotarsus," Trans. S. Afr. Phil. Soc. **16**: 201-204. 1906.

E. Fraas: "Die neuesten Dinosaurierfunde in der schwabischen Trias," Die Naturwissenschaften **1**(45): 1097-1100. 1913.

F. von Huene: "Die Dinosaurier der europäischen Triasformation." Geol. u. Pal. Abh., Supplem. **1**: 1908.

———: "Ein primitiver Dinosaurier aus Elgin," Geol. u. Pal. Abh. **14** (N. S. **10**) (**1**): 1910.

———: "Beiträge zur Geschichte der Archosaurier," *ibid.* **17** (N. S. **13**) (**1**): 1914.

———: "Ueber die Zweistämmigkeit der Dinosaurier," Neues Jahrb. Beil. **37**: 577-589. 1914.

F. von Huene & R. S. Lull: "Neubeschreibung des Originals von *Nanosaurus agilis* Marsh," Neues Jahrb. **1**: 134-144. 1908.

——— ———: "On the Triassic Reptile *Hallopus victor* Marsh," Amer. Jour. Sci. **25**: 113-118. 1908.

O. Jaekel: "Ueber die Wirbethierfunde in der Oberen Trias von Halberstadt," Palæont. Zeitsch. **1**: 155. 1913.

R. S. Lull: "Fossil Footprints of the Jura-Trias of North America," Mem. Boston Soc. Nat. Hist. **5**: 461-557. 1904.

———: "Dinosaurian Distribution," Am. Jour. Sci. **29**: 1-39. 1910.

———: "The Life of the Connecticut Trias," *ibid.* **33**: 397-422. 1912.

O. C. Marsh: "Notes on Triassic Dinosauria," *ibid.* **43**: 543-546. 1892.

———: "Restoration of Anchisaurus," *ibid.* **45**: 169-170. 1893.

———: "Dinosaurs of North America," U. S. Geol. Sur. 16th Annual Report, 143-244. 1896.

99. R. S. Lull: *l. c.*, 482. 1904.

From these and other fragments of evidence, we may reconstruct a concept of the dinosaurs as a land adaptation developed during the arid Permo-Triassic climatic phase, corresponding to the later deployments of the mammals along the same lines of adaptation and under a similar impelling cause of progressive aridity and continental expansion. During the base-leveling and submergence and moist tropical climate of the Jura, these dry-land adaptations reinvaded the swamps and coast-marshes, the least specialized types (*cf. Protorosaurus*), more quadrupedal and some of them long-necked, reverting farthest towards an aquatic life and specializing into the peculiar Sauropoda, while the higher bipedal types retained more of their terrestrial habitat but evolved into huge, massive armored and bizarre creatures, to be paralleled in habit and type at a later date by the bizarre specializations of the Eocene Mammalia. These are the familiar dinosaur fauna of the Upper Jura and basal Cretaceous. The drier uplands of that time must have been tenanted by lighter, smaller dinosaurs, but of these, in my opinion, we have little direct evidence. But that they continued to exist and carry forward their primary lines of adaptation is shown by the subsequent history of the order.[100]

In the Lower Cretaceous occurred a swing towards emergence and arid conditions, not extreme, but sufficient to wipe out the sauropod dinosaurs in the Northern world. They survived, however, in the southern continents until, in the middle and later Cretaceous, the pendulum swung back to a marked extreme of submergence and moist-tropical climate, and their remains are found in late Cretaceous beds in South America, East Africa, Madagascar and Australia. The correlation of these beds is in need of revision, however; they may be Comanchean. In the Northern world, at all events, they did not reappear after the early Comanchean. A dinosaur fauna largely similar to that of the Jurassic in habits and adaptation in other respects, developed during the late Cretaceous in the North. It contains no Sauropoda, but it includes amphibious types (Trachodontidæ) with marked aquatic adaptation, gigantic terrestrial swamp and forest dwellers, like the ceratopsians, tyrannosaurs and ankylosaurs, and many smaller more agile forms. These Cretaceous giants, however, appear to have evolved, not from amphibious or aquatic dinosaurs of the Jura, but, in part at least, from small and little known forms, of more upland adaptation, which had been much more highly specialized for dry-land life

100. R. S. Lull ("Dinosaurian Distribution," Amer. Jour. Sci. **29**: 1-39. 1910) has admirably summed up the data regarding the geological occurrence of dinosaurs. While not agreeing in all respects with his interpretation, I take pleasure in noting the accuracy and clear presentation of the evidence as worthy of the high regard in which its author is held by his confrères.

than any of the Jurassic swamp dwellers, and had readapted themselves to the forest and swamp environment of the later Cretaceous. The trachodonts and ceratopsians, for instance, while related to the earlier iguanodonts, cannot be directly derived from them but must be traced back to some unknown contemporary which was highly progressive in developing efficient grinding dentition, compact feet with flattened hoofs, etc.—characters which in a survey of mammalian adaptation we find to be especially associated with upland habitat. The evidences of former dry-land adaptation are not so clearly shown in the other swamp-giants of the late Cretaceous, but they may perhaps be shown by further study.[101]

In sum, we may find in the hypothesis of recurrent climatic change, and in the primary adaptation of the dinosaurs as a dry-land adaptation of Reptilia and their secondary readaptations to forest and swamp life, a fairly satisfactory solution of their distribution and phylogeny. Lull, in his able discussion of the subject (1910), explains their adaptation along these lines. But at present our data, both of correlation and identification, are too uncertain to allow of positive and detailed conclusions in regard to the centers of dispersal and course of migration of the dinosaurs. That the sauropods survived in the southern continents long after their extinction in the north appears proven, if we accept the stated geological correlations of the southern formations where they are found and set aside as an erroneous identification the reported occurrence of a sauropod in the Danian of France.[102] That the Theropoda survived into the Eocene in South America and Theropoda and Predentata into the Paleocene in North America is not improbable on *a priori* grounds, but the evidence that they actually did so survive is open to serious question. So far as they go, the facts accord with the dispersal of the dinosaurs from the northern land mass. And so far as I have been able to review the data, the migrations of the order could be made to conform with the present distribution of continental and abyssal areas (Madagascar excepted[103]) about as well as with the different distribution upon which they are plotted by Dr. Lull.

It is significant in this connection to note that young individuals are very rarely found in the dinosaur formations. Thousands of individuals

---

101. L. Dollo (Bull. Soc. Belg. Géol. **19**: 441. 1905) has shown that the quadrupedal gait of many of the Predentate dinosaurs is a secondary adaptation from bipedal ancestry. I believe this to be true, to a less extent, of the Sauropoda as well.

102. F. Nopsca (Rep. Geol. Mag. **7**: 261. 1910) states that the femur on which this recorded occurrence is based is not a sauropod but a trachodont dinosaur, allied to or identical with Telmatosaurus of the Gosau beds of Austria.

103. The Cretaceous Sauropoda of Madagascar may have reached that island in the same manner as the hippopotamus did at a later period, namely by swimming.

are found together in some of the great quarries, pertaining to a great number and variety of genera and with a wide range in size, but it is very rare to find young individuals among them. This fact is well known to collectors, but has not, as far as I know, been commented upon in print. It is true that young individuals are less clearly distinguished from adult among reptiles than among mammals, the chief difference being the imperfect ossification of the bone structure, and that such imperfectly ossified bones are likely to be poorly preserved and might often be rejected by collectors on this account. But making all reasonable allowance for these considerations, there remains a very notable contrast with fossil mammal quarries and fossiliferous formations, in which young individuals are always to be found among any considerable number of adult specimens and often are more numerous than mature individuals.

This may be interpreted in conformity with the above theories as to the habitat of dinosaurs, by supposing that the young dinosaurs were more dry land or upland animals, retaining the ancestral habitat, and coming down into the swamps only when they reached maturity and their larger size made an amphibious or aquatic habitat more suitable. The young animals would rarely or never visit the swamps and deltas, whose formations have alone been preserved, and their fossil remains would be correspondingly scarce.

Young crocodiles, so far as I can gather from various descriptions, are somewhat more terrestrial in habit than the full-grown animal, but the difference is evidently not considerable. Analogous cases among fish, marine types breeding in fresh water and vice versa, are well known. The migration of birds has also some analogy, if, as may often have been the case, the swamp dinosaurs resorted to dry land for breeding and egg-laying purposes. In either case the breeding or egg-laying place would be presumptively the ancestral habitat of the race.

Chelonia

The publication of Dr. Hay's splendid monograph[104] upon the extinct Chelonia of North America has added a great deal to the available data for explaining the distribution of this group. So far as the Tertiary and modern distribution goes, it conforms to the same lines of dispersal as do the various orders of mammals. The pre-Tertiary history of the order is mostly too fragmentary to afford any important data bearing, pro or con,

104. O. P. HAY: "Fossil Turtles of North America," Carnegie Institution Publ. **75**: 1908.

upon the theories here presented. The whole order is in general conservative and persistent to a high degree, like the Crocodilia.

The occurrence of giant tortoises (*Testudo*) on several oceanic islands and in Australia and Patagonia (*Meiolania*) has been adduced as evidence for continental connection of these islands and for an Antarctic connection of the two southern continents. Here, as in the case of the carnivorous marsupials cited on page 96, the evidence will not bear close examination. In the first place, we know that large tortoises of the genera *Testudo* and *Stylemys* are among the most abundant fossils in the Middle and later Tertiary of the Nearctic, Palæarctic, Oriental and Ethiopian regions. So far as we can judge, they were cosmopolitan, except Australia and Patagonia. They occur in the Pleistocene of Cuba and Madagascar and survive to the present day in certain islands in the Indian Ocean and in the Galapagos Islands. So far as these oceanic islands are concerned, if we assume that their presence in one involves continental union, it must do so in all. If such continental union occurred, it is hardly conceivable that, in each instance, tortoises alone would have made their way to the islands. We must infer for each and every one of them a vertebrate and invertebrate land fauna. Where is that land fauna, and why has it perished? The idea of selective drowning might possibly be entertained if we had to do with only a single instance, but is too absurd for serious consideration, when we deal with several instances of the survival of the same race. The only reasonable method of accounting for the presence of *Testudo* on these islands is that its facilities for oceanic distribution are somewhat better than those of mammals and that it arrived by over-sea transportation.

The most recent argument for land connection of the Galapagos Islands is by Dr. Hay.[105] He advocates a connection with Central America, via a submerged ridge which is shown in the reports of the Blake Expedition to extend southwest from Costa Rica towards the islands. The depth of this ridge Dr. Hay omits to state, but the soundings indicate it as being upwards of ten thousand feet, so that it does not materially affect the improbability of an elevation to this extent. The Galapagos Islands are purely volcanic in origin and stand upon a platform less than a thousand feet in depth, similar on a smaller scale to that which surrounds the continents and presumably open to similar interpretation. If so, the islands have, probably, been more or less completely united at periods of continental emergence and completely isolated at periods of continental sub-

---

105. O. P. HAY, *l. c.*

mergence (if any such have occurred since they were first upbuilt from the ocean floor by volcanic ejectamenta) but never connected with the mainland. As the island platform is less extensive than Madagascar or Cuba, farther from the mainland and without intervening island stepping-stones, the opportunities for successful colonization through rafts or other means of transport have been fewer, and have not succeeded in introducing any mammals or amphibians and but few reptiles and invertebrates. The most favorable opportunity for such colonization would be when the islands were at their maximum elevation—towards the end of the Tertiary, if this corresponded with the elevation of the mainland—as at that time the extent of coast and consequent probability of making a landing would be much greater. The subsequent isolation of the islands by submergence accounts for the presence of distinct although related species on different islands. Thus the series of "miracles of transportation," which Dr. Hay finds it so difficult to accept, dwindles down to a single "miracle" and to one which he must invoke to account for the populating of the more remote Pacific islands, and which, when considered in relation to the time involved, does not really involve any serious improbability. On the other hand, if a miracle be an exceptional occurrence in apparent contravention of all probabilities, and without assignable causes in natural law, I think the processes of selective drowning, or of selective migration of sporadic elements of a fauna, involved in the alternate hypothesis, in addition to the elevation during the late Tertiary of abyssal depths to the surface, unwarranted by any valid evidence, does involve a series of miracles, almost as unworthy of belief on the evidence offered, as the special creation of the species of the Galapagos Islands appeared to Darwin.

The present distribution of species of *Testudo* on the islands of the Indian Ocean has been partly changed by man, so that there is some uncertainty about its details. Lydekker states it as follows:

> "Madagascar, probably the Comoros, North and South Aldabra—small islands lying to the northwest of the northern point of Madagascar—the Mascarenes or Mascarenhas, situated to the east of Madagascar and including Réunion, Mauritius and Rodriguez and lastly the Amirantes and the Seychelles, which are the most northern of the whole assemblage and only about four degrees south of the equator."[106]

Each of these groups of islands, except the Mascarenes, stands upon a shallow platform, and is surrounded by abyssal ocean, upwards of 5000 feet between the Comoros and Africa, elsewhere upwards of ten thousand feet. The three Mascarene islands rise separately from abyssal depths. Madagascar is about 180 miles from the African coast; the other islands

106. Science Progress 303. October, 1910.

are 400 to 600 miles from Madagascar; the present normal set of current is unfavorable to transportation from Madagascar.

It is very frequently asserted that a bank of shallow soundings connects India with Madagascar through the Amirante Seychelles group, and that this indicates a former continental bridge of which these islands are remnants. The facts are as above stated; the so-called bank is very little above the general level of the floor of the Indian Ocean and is not differentiated from it in any features of relief that would suggest its former continental character.

The transportation of natural rafts five hundred miles against the normal set of current—or five times that distance if from the East Indies—is the most improbable element in this explanation. There is no valid reason to suppose that the general direction of winds and currents differed materially in the later Tertiary from the present day conditions. I do not think it necessary to assume with Dr. Lydekker that the tortoises were of gigantic size when they reached the islands or to ignore, as he does, the elements of parallelism in considering their affinities to continental species. Nor does it appear that the difficulties which he admits in accounting, on the hypothesis of former continental union, for the absence of the rest of the fauna, should be "set aside for future consideration." They add so greatly to the improbability of the hypothesis, that in conjunction with the physiographic difficulties it appears wholly out of range of reasonable probability. On the other hand, an investigation of the very variable direction of the winds and currents in the Indian Ocean would probably yield data to reduce the improbabilities in the hypothesis of over-sea transportation as above stated. The third possible hypothesis is that the present distribution is due in part to human agency, not necessarily limited to the historic period. If this factor may account for a species of *Canis* in Australia distinct from the living species of Arctogæa, it may perhaps help to account for peculiar species of tortoises as well.

As for *Miolania,* it occurs in the *Notostylops* Beds of Patagonia and in the Pleistocene of Australia. The *Notostylops* Beds are Eocene, as here advocated. The persistence of so highly specialized a genus for so long a period appears surprising; if they are Lower Cretaceous, as Ameghino asserts, it is quite unprecedented. My acquaintance with chelonian anatomy is not adequate to warrant my venturing an opinion as to how far parallel evolution from less specialized Pleurodira might account for this anomaly. But we certainly do not know to what extent this genus or a less specialized pleurodiran ancestor may have been aquatic or even marine in its habits. And unless we suppose that it had some such semi-marine adaptations which would enable it to cross a marine barrier impossible for terrestrial mammals, I do not see how to account for its reaching Australia without any of the *Notostylops* mammalian fauna accompanying it. We cannot believe that a placental fauna ever reached Australia, for if it had we should not see the development of a marsupial fauna on analogous adaptive lines to take its place. *Miolania,* then, could cross some barrier, presumably an ocean barrier, which land mammals could not; and it becomes merely a question of how wide a barrier this extinct chelonian of unknown habits could cross. The present lines of the continents within the continental shelf would not present materially greater difficulties in its reaching Australia via Antarctica than *Testudo* has managed to surmount in reaching Mauritius and the Seychelles, and I think we are justified in saying that the occurrence of *Miolania* has no weight as evidence of former Antarctic connections of the Southern continents and, in fact, is opposed to any actual land connection.

The following notes on the distribution of the land Chelonia are summarized from Dr. Hay's monograph:

Crypotodira are the dominant group of turtles and compare with the placentals among mammals. All continents except Australia.

Chelydridæ.—Central America, eastern North America and New Guinea. Apparently a relict-distribution, but the family is unknown fossil.

Dermatemydidæ.—Part of Central America. Found in abundance in North America in the Upper Cretaceous and in reduced numbers during the Tertiary.

Emydidæ.—Chiefly Holarctic and Oriental. A few have reached South America, none in Ethiopia, Madagascar or Australia. First known in Holarctic Lower Eocene.

Testudinidæ.—Very abundant in Tertiary Holarctica but now mostly restricted to its southern margin. Abundant now in Ethiopia and a few species in Neotropical and Oriental regions also in oceanic islands. Present in Sumatra, absent in Java, present in Celebes but absent in Borneo. These and other features are very suggestive of man's having had much to do with the local extinction of Tortoises. For obvious reasons this family would be peculiarly subject to his ravages.

Pleurodira.—Now limited to the three southern continents, Holarctic in the later Mesozoic and early Tertiary, and the extinct Amphichelydia from which they are descended were likewise a Holarctic group. The occurrence of closely related genera in South America and Madagascar is used in support of a Brazilian-Ethiopian-Malagasy land connection. It would be interpreted in conformity with the views here advocated, as due to common descent or to parallel evolution from Tertiary Pleurodira of Holarctica.

Trionychidæ.—The distribution of this group is exceptional in that it is entirely absent from the Neotropical region and the Pacific coast of North America, while common to eastern North America, the Ethiopian, Oriental and southeastern Palæarctic regions and New Guinea. Ameghino records *Trionyx* from the *Notostylops* Beds of Patagonia,[107] indicating if the identification be correct that the group was formerly present in South America. It is found abundantly in the Cretaceous and Tertiary of North America and in the older Tertiary of Europe; absent from Australia and Madagascar.

---

107. FL. AMEGHINO: "Age des Formations Sédimentaires de Patagonie," Anal. Soc. Cient. Argent. **50, 54**: 52 of separata. 1903.

Presumably this is a relict-distribution of an ancient group, whose facilities for transportation were relatively limited. It should be noted that the hypothesis of over-sea transportation on rafts would be less applicable to aquatic animals than to their terrestrial relatives, as they would be less likely to be carried out to sea on floating vegetation, on account of their ability to leave it at will for the shore. But the absence of the group from the Neotropical and Western Nearctic, and its presence in New Guinea, are anomalous features.

Crocodilia

The crocodiles are usually regarded as the most conservative of the reptilian orders. This is true enough, so far as adaptive specialization from the primitive amphibious environment into the higher plane of terrestrial habitat is concerned. Their expansional tendencies have been in the other direction, towards invasion of the marine province.

The present geographic distribution of the group is as follows:

Narrow snouted	*Gavialis,*	India.
	Tomistoma,	East Indies.
Broad snouted	*Alligator,*	Southern United States, China.
	Crocodilus,	Africa, southwest Asia, Oriental and northern Australian regions, tropical America and West Indies.
	Caiman,	Tropical America.
	Osteolæmus,	West Africa.

This is very clearly a remnant-distribution and is explained, at least in part, by the occurrence of crocodiles in the Tertiary. Fossil Crocodilia are abundant in the early Tertiaries of Europe and North America. The European species, according to Zittel,[108] belong partly to *Crocodilus,* partly to the extinct genus *Diplocynodon,* in which the proportions of the skull are like *Alligator,* but with a notch for the lower canine, like *Crocodilus,* while the armor, especially the belly armor, is like that of *Caiman.* North American Tertiary Crocodilia are all with one exception referred to *Crocodilus,* but the armor is incompletely known, and they may prove also to include *Diplocynodon. Gavialis* is recorded from the late Tertiary of India; *Tomistoma* and *Crocodilus* occur in the Oligocene of Egypt and *Tomistoma* in the Miocene of southern Europe. The common Egyptian Oligocene species of *Tomistoma* is intermediate between this genus and *Gavialis.*

The Upper Cretaceous crocodiles are nearly allied to those of the early Tertiary.

The Jurassic and Comanchic crocodiles include also long-snouted gavial-like forms, more or less marine in habitat, and broader-snouted crocodile

108. K. A. von Zittel: Grundzüge der Palæontologie, 2e Aufl., ii Abteil: 272. 1911.

or alligator-like forms of more strictly fresh-water habitat. All have amphicœlous vertebræ, whereas all Cenozoic and late Cretacic crocodiles have procœlous vertebræ. It is commonly believed that certain of the narrow-snouted types (*Teleosaurus*) led into the gavials, the broad-snouted (*Goniopholids*) into crocodiles and alligators; and that the broad-snouted types first appearing in the Upper Jura are derived from the teleosaurs

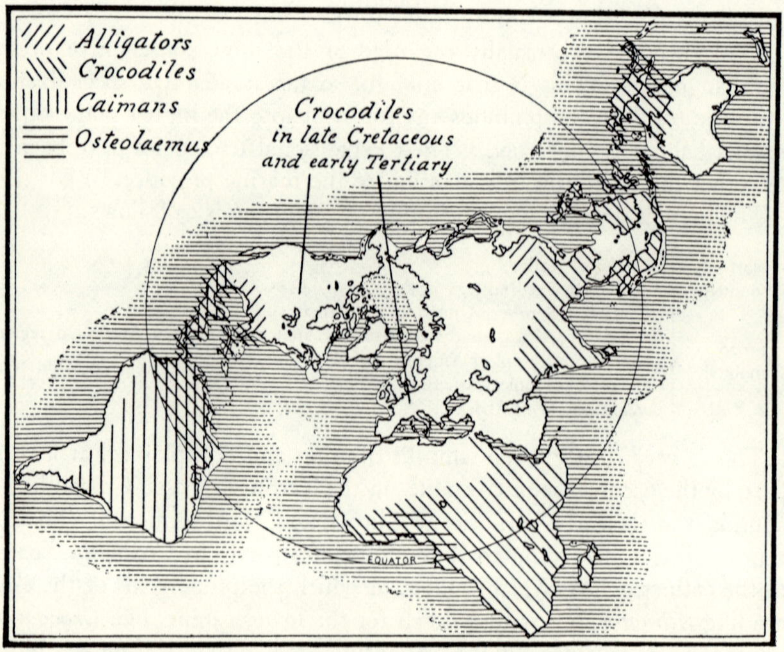

FIGURE 32. Distribution of the Crocodilidæ

Originating probably in Cretaceous Holarctica, they have been restricted to the peripheral continents by inability to become adapted to cold climates. Note discontinuous occurrence of crocodiles and of alligators, the last the most specialized, as *Caiman* and *Osteolæmus* are the most primitive of the living genera.

which first appear in the Middle Jura. Neither of these propositions seems to me to be probable. The narrow-snouted crocodiles are characteristic of marine or semi-marine formations, the broad-snouted kinds of fresh-water formations; the known formations of the Middle Jurassic are chiefly marine, those of the Upper Jura chiefly fresh-water.

If we turn back to the Trias, we find that in the allied Parasuchia there were also long-snouted (*Mystriosuchus* and *Rutiodon*) and broader-snouted (*Belodon*) types—both of fresh-water habitat, but apparently less aquatic than *Crocodilia;* in the allied Pseudosuchia the snout was short, and the adaptation to amphibious or fresh-water life; while the more distantly related dinosaurs were terrestrial and short-snouted. Upon these data, it appears to me more reasonable to suppose that the Triassic *Mystriosuchus* and *Rutiodon,* the Jurassic Geosauridæ, Teleosauridæ and Metriorhynchidæ and the Tertiary Gavialidæ are all independent successive adaptations to a fish-eating diet and a more or less marine habitat and that the Jurassic Goniopholidæ are the source of all the modern Crocodilia. This will also relieve us from the necessity of supposing that procœlous vertebræ and a number of other identical characters were independently and simultaneously acquired in two phyla of diverging adaptation. The accepted view involves the anomaly of associating divergent adaptation with convergent structural evolution.

However this may be, we are justified in assuming certain characters as primitive among the modern Crocodilia, since they are common to all the older types. These are the following:

(1) More complete and consolidated ventral armature. Common to all the Mesozoic genera, retained in *Diplocynodon* of the European Tertiary and the modern *Caiman* and *Osteolæmus.*

(2) A notch instead of a pit in the upper jaw for reception of the lower canine. Common to all the short-snouted crocodiles of the Mesozoic and Tertiary, retained in the modern *Crocodilus.*

(3) Amphicœlous vertebræ. Common to all Crocodilia and related groups up to the middle Cretaceous, lost in most Upper Cretaceous and all Tertiary[109] and modern genera.

(4) Large supratemporal and small lateral temporal fenestræ. The upper temporal fenestra is large in all Mesozoic Crocodilia, considerably small in the gavials, quite small in *Crocodilus, Alligator* and *Caiman.*

(5) Posterior nares more anterior in position. In the Mesozoic crocodiles, the Choanæ are situated at the posterior end of the palatines in the long-snouted groups, while in the short-snouted Goniopholidæ, they have moved further backward, between the palatines and pterygoids. In the modern Crocodilia, they are still further backward, entirely enclosed within the pterygoids. This is an adaptation to lying submerged with the nostrils only projecting above the surface of the water and enables the animal to breathe comfortably in this position. It would naturally develop in the slow, omnivorous broad-snouted crocodiles and not in the swift-moving fish-catching, long-snouted types; hence its greater development in Goniopholidæ than in teleosaurs, etc. The fact that it is fully as much developed in gavials as in crocodiles is another reason for deriving both from Goniopholid ancestry.

According to the above criteria, *Alligator* is the most progressive modern genus.[110] *Caiman* is primitive in (1); *Osteolæmus* in (1) and (2);

---

109. Except *Notosuchus* of the Patagonian Eocene.

110. R. L. Ditmars, of the New York Zoölogical Park, has observed that crocodiles are decidedly more active and ferocious animals than alligators. I would not interpret this, however, as meaning that they are more progressive, in the sense here used, since the adaptation of the typical Crocodilia is not towards an active life.

Crocodilus in (2) and to some extent in (4); *Gavialis* and *Tomistoma* are primitive in (4), divergent in adaptation in other respects, so that comparisons would be unprofitable. We may conclude, therefore, that so far as they go, the Crocodilia accord with the general lines of distribution of other groups. They ranged much farther north during the Tertiary than they do now; the most progressive modern genus, *Alligator,* has the most northerly range, and the Neotropical *Caiman,* the West African *Osteolæmus* and the cosmopolitan tropical genus *Crocodilus* are primitive in one or another respect. The gavials also had a wider and more northerly distribution during the Tertiary.

That the present limits of range are conditioned chiefly by temperature and climate, and that the much wider range in the early Tertiary was due to a warmer climate towards the poles, will hardly be questioned. Of previous limitations and expansions of range in the order, due to previous secular alternations of climate, there is no adequate evidence. The distribution of the more primitive modern genera in widely separated parts of the tropics; the occurrence of the most progressive genus on the northern borders of the range of the order in two widely separated regions, and, finally, the survival in the Eocene of Patagonia of a crocodile, *Notosuchus,* of the Mesozoic type which had disappeared from the Northern world by the Middle Cretaceous,—these facts point to a northern rather than a tropical or southern center of dispersal for the order; but the evidence is slight and far from conclusive.

Lacertilia

Lizards are the largest group of the Reptilia, comprising over 1800 species, mostly of small size. Most of them are active animals, and a large proportion are adapted to rocky and desert habitat and arid climate. They are more dependent on external warmth than mammals and birds, and consequently are excluded from the colder regions; their means of dispersal are perhaps less limited than with mammals, if we may judge from their wider distribution, for they do not appear to be of more ancient origin. Unfortunately, the rarity and fragmentary nature of their fossil remains stands in marked contrast with those of mammals, and our evidence as to their evolution and dispersal is chiefly indirect, based upon the modern distribution, and is neither conclusive nor convincing. Such as it is, it compares fairly well with corresponding distribution features among the smaller Mammalia and points to the same conclusions. But it emphasizes the importance of occasional over-sea transportation as a factor in distribution. Gadow observes[111] in regard to the Geckos, the most cosmopolitan of all lizards:

> "Although not at all aquatic, they are particularly fit to be transported accidentally on or in the trunks of floating trees, to which they cling firmly, and they can exist without food for months."

Other groups are somewhat less easily transported in this way, and to quote the same authority:

> "It is a most suggestive fact that most of those families of Reptiles, and even of other vertebrates which have a wide distribution and are apparently debarred from transgressing Wallace's line, are also absent from Madagascar."

The iguanas are chiefly Neotropical, but they occur also in Madagascar, in the Fiji and Friendly Islands and in the West Indies and Galapagos Islands, as well as on the American continent. Fossil iguanas are recorded from the Upper Eocene and Oligocene of Europe and from the Upper Cretaceous and Middle Eocene of the western States. If these determinations be correct, they must formerly have been more cosmopolitan. Their presence in Madagascar is most reasonably explained by their former presence in Africa, which is rendered probable by the fact that they occur in the early Tertiary of Holarctica, along with various mammalian groups which certainly did reach Africa. Their disappearance from the mainland of Africa may be coupled with the invasion of other later developed groups, Zonuridæ, Varanidæ, Lacertidæ, which were prevented from reaching the New World by the absence of any land bridge or land approximation within their temperature limits. One genus of Zonuridæ has likewise reached Madagascar.

Bearing in mind the progressive limitation of northerly range of the Lacertilia by the secular refrigeration of the polar regions during the Cenozoic, we can see that, if the distribution of land and water has not greatly changed except within the 600 feet limit, any families arising during the middle or later Tertiary would be limited to the Old or to the New World. While the distribution of various lizards in oceanic islands compels us to admit that they can cross considerable bodies of water and obtain a foothold on an imperfectly populated island area, yet the probabilities of their crossing the whole width of a broad ocean and maintaining themselves against competitors trained in the broad arena of a great continent appear to be very much less and almost negligible. Conversely then, we may assume that a distribution, such as that of the Scincidæ,

111. Hans Gadow: Cambridge Natural History **8**: Amphibia and Reptiles. 1901. The distribution data for lizards and amphibians are mostly based upon this authority.

Iguanidæ, Geckonidæ, Anguidæ and Amphisbænidæ, involves the evolution and cosmopolitan distribution of these families as early as the Eocene. The Agamidæ, Varanidæ, Lacertidæ, Zonuridæ, Chamæleontidæ are Old World families, and none are known from the New World. The Zonuridæ may well be regarded as of Ethiopian evolution; if not, they must be a remnant of a very ancient stock. The same may be said of the Chamæleons, except that if Ethiopian they reached as far as India. The Lacertidæ, the highest, or at least most typical family of lizards, are evidently the most recent development; they have not yet reached Madagascar or Australia, and their northern limit is higher than in any other lizards. The Varanidæ and Agamidæ have not reached Madagascar but have spread widely through Australia. The evidence from extinct lizards is very slight, the remains are scanty and mostly too fragmentary for positive family identification. Of the several genera from the Eocene and Oligocene of North America, two are positively referable to the wormlike Amphisbænidæ, whose present distribution in tropical America, the West Indies and Africa is thus partly explained as a remnant of a former wider northerly range and presumably Holarctic. Of the remaining North American Tertiary genera, *Peltosaurus* and *Glyptosaurus* are referred to the Anguidæ;[112] the remaining genera are too fragmentary for reference or have not been studied.[113]

In general, so far as I can judge, the Lacertilia lend no support to the theories of transoceanic bridges. Their widespread insular distribution must in some cases, and may in most others, be explained by over-sea transportation. They lend some support to late Tertiary elevations to the continental shelf line so as to include the continental islands and to a line of separation in the East Indies which some, but not all, were able to cross; those which did succeed in crossing it spread widely through Australia, indicating more continental conditions, and also indicating in these families a capacity for crossing marine barriers which enabled some of them to reach Madagascar, New Zealand and various Pacific islands.

The ratio of their abundance in regional faunæ is apt to be inversely to the full development of mammalian life. Where mammals are scanty, as in oceanic islands, lizards partly take their place; and this is true of

---

112. EARL DOUGLASS: Ann. Carn. Mus. 4: 278. 1908.

113. The recorded presence of Iguanidæ (*Iguanavus*) in the Cretaceous and Eocene, while not provable, is not unlikely; that of *Chamæleon* (*C. pristinus*) in the Upper Cretaceous is improbable and based upon insufficient evidence; the reference of *Thinosaurus* (Middle Eocene) to the Varanidæ appears to be merely a matter of bibliographic convenience; the specimens are probably definitely referable, but the only expressed opinion as to their affinities is by Boulenger (1891), who suggests their relationship to the Teiidæ.

some continental regions as well as of oceanic islands. In the typical continental fauna, the lizards are largely restricted to desert or rocky habitat and are of small size. Yet these last are the most typical members of the order. They show what its primary adaptation was. Various readaptations appear, to fossorial, to aquatic, to arboreal or to terrestrial forest life, repeated again and again in different families and causing frequent parallel divergencies from the primary type. This primary type, I regard as an adaptation to a Mesozoic arid period. The moist uniform climatic phase of the early Tertiary would tend to develop large forest living and aquatic forms and restrict and provincialize the more typical lizards. During the middle and later Tertiary, the typical lizards would expand and multiply in numbers and variety, but, on account of their lack of adaptability to cold climate, their evolution was not so much a successive series of dispersals from a Holarctic center, as a provincial evolution from the arid centers of the great continents. Such *a priori* hypotheses are of little value, however, except as confirmed, modified or refuted by detailed study of the affinities and geographic distribution of the genera of each family, checked by a wider knowledge and more thorough study of the fossil forms. Until the fossil Lacertilia have been thoroughly studied and their affinities authoritatively estimated, any conclusion whatsoever as to the evolution and distribution of the order remains highly hypothetical.

Dr. Gadow's recent study[114] of the distribution of *Cnemidophorus* and its interpretation is an excellent example both of the value of such detailed studies and the need of carefully distinguishing between what the data themselves indicate and what is assumed as true from other evidence. He concludes,—

(1) That the species are the product of their environmental conditions;

(2) That their dispersal center was in western Mexico, whence they have spread northeast as far as Texas and Florida, southwardly into South America, northwestwardly into Lower California;

(3) That the primitive type was nearest the Texas and Florida species.

He assumes—evidently on some other grounds—

(1) That a great land area stretched out from Mexico far into the Pacific during the Tertiary all the way between Lower California and Central America;

(2) That the central tableland of Mexico was a vast fresh-water lake during most of the Tertiary;

(3) That Cuba was connected with the American mainland during the Oligocene (this assumption underlies the statement that, since the Floridian *Cnemidophorus* did not reach Cuba, its migration must have occurred as late as Miocene).

---

114. H. Gadow: "A Contribution to the Study of Evolution based upon the Mexican Species of *Cnemidophorus*," Proc. Zool. Soc. London 1: 277-375. 1906.

Ortmann,[115] reviewing this paper, takes, as proven by Gadow's studies, not merely the points actually indicated but also the assumptions which are entirely unnecessary to explain the data but which Dr. Gadow evidently feels obliged to take for granted. In fact, these assumptions interfere with a reasonable interpretation rather than help it, and all of them are questionable, to say the least. The great Tertiary lake is, I suspect, on all fours with the vast interior "lakes" of the Plains region of the United States, which the progress of physiographic and paleontologic studies have relegated to the domain of myth. The connection of Cuba with the mainland of either North or South America involves the same difficulties as the connection of Madagascar with Africa. The recent discoveries by Dr. de La Torre of a Pleistocene vertebrate fauna in Cuba strongly confirm this analogy between the Cuban and Malagasy faunæ. The existence of extensive land west of the present Pacific coast line is an equally unnecessary and improbable hypothesis. On the other hand, Dr. Gadow fails to take into account the barrier between North and South America which prevented or hindered intercommunication of land faunæ during a large part of the Tertiary, while it permitted intercommunication of marine faunæ during the Eocene. I am not here concerned with its nature but may venture to point out that its bearing on the differentiation of species would be important. For, once across that barrier, an invading species would find itself in unfamiliar environment on account of differences in the autochthonic fauna and flora, even though the physical environment were similar. If the rising of the Mexican tableland conditioned the dispersal of the genus from that center, we can see in this different biotic environment the reason why the marginal species in North America should be primitive, while the marginal species of South America are highly specialized. In general, it would be true that the species of the dispersal center (or those nearest to it, where, as in this case, it has become ill adapted for the habitat of the race) will be the most progressive and those of the marginal areas nearest the primitive stock. But where the scattering primitive forms, in following the primitive climatic conditions, are brought into a new floral and faunal environment, this may profoundly modify them and cause a rapid divergence and specialization.

DISPERSAL OF BIRDS

As a class, birds are extremely difficult in their taxonomy. They are held closely to type in comparison with mammals, and the differences be-

115. A. E. Ortmann: Geog. Jahrb. 31: 262. 1908.

tween them are mostly directly and obviously due to adaptation. Adaptive parallelism obscures the true affinities to such an extent that even at the present day the major classification is somewhat uncertain. This difficulty is the greater on account of their rarity as fossils. There is no reason to interpret this rarity as indicating any lack of abundance of birds in the faunæ of Tertiary and later Mesozoic time; it is presumably to be accounted for by their generally upland habit, small size and the lightness and fragility of the skeleton. The small minority of fossil birds which are known from anything more than a few fragments are, with two or three notable exceptions, aberrant types—ground-birds, marine or lacustrine types, whose habitat facilitated their preservation as fossils. By far the most notable and instructive of these exceptions is *Archæopteryx*.((115a))

It has been customary to class the greater number of the ground-birds (Ratitæ) as a more primitive sub-class. On *a priori* grounds, this may be correct enough, since it would appear theoretically that feathers must have preceded flight, the ability to fly being conditioned by high organization plus small size, and this would involve a rapid circulation and high temperature, which could hardly be attained without a non-conducting coating over the body. But it appears certain that most, and possible that all of the existing ground-birds are readaptations to terrestrial habitat from flying ancestors, and their resemblances are due almost wholly to adaptive parallelism.

Owing to their powers of flight, the dispersal of birds is much less limited and conditioned by distribution of land and water or by mountain or desert barriers than is that of mammals. Climate and environment are much more important factors. Their dispersal is accordingly much wider, and this is especially true of the more migratory and strong-flying types. The general course of their dispersal from the northern land masses is in some respects much more obvious than with the Mammalia, provided we allow for the extreme imperfection of their geological record; but on this account, it is not supported by the mass of direct evidence which we have among mammals.

The most primitive living birds, the penguins, are Antarctic in their distribution, and as fossils are known only from the Antarctic Tertiaries, where they include gigantic terrestrial adaptations. It is of interest to note that the only actually known land vertebrates of the Antarctic continental area are penguins. If this continent had been united during the late Mesozoic and early Tertiary to Australia and South America, we

((115a. Fossil Birds. See Naturwiss. Woch. **1918**: 352-364, review by K. Lambrecht of the known records.))

should expect to find a fossil mammal fauna, probably highly progressive and specialized before the spreading ice swept it out of existence. We might, indeed, hope to find a few marine adaptations from this mammalian fauna still haunting the edges of the Antarctic pack. But in fact, the three items which to my mind have a bearing upon early Tertiary conditions in Antarctica all point towards continued isolation and obviously parallel the fauna of oceanic islands. These are,—

(1) Gigantic land-penguins in the ? Eocene deposits of Seymour Island (also in Patagonia). Compare with the gigantic land birds of various oceanic islands, correlated with paucity or absence of land mammals.

(2) The living marine penguins are not readily interpreted as a primarily marine adaptation, but they are very easy to understand as modified survivors of a group formerly of terrestrial habits, altered to meet the present conditions under which alone could life be maintained on the Antarctic shores.

(3) The occurrence of *Miolania,* as interpreted on page 113, is suggestive of the former presence of giant land-turtles in Antarctica, although not explainable as evidence of former land connections with South America and Australia.

There may be other indirect evidence in the distribution of marine Vertebrata and Invertebrata, which, if conservatively interpreted, would confirm or disprove these indications. So far as they go, they suggest that ground-birds and land-turtles were the large land vertebrates of Tertiary Antarctica as in oceanic island faunæ of to-day.

The distribution of modern land birds is universally interpreted in terms of Northern derivation. Oceanic, desert or mountain barriers have been much less efficient in limiting their range, and the efficiency of the climatic factor is much more obvious than with mammals. Their dispersal from a Holarctic center in successive waves of migration is indicated by the dominantly Holarctic habitat of the highest and latest developed groups, by the generally tropical habitat of archaic groups often highly specialized, whose ancestors or relatives are in many cases known from the Holarctic Tertiary, and by the fact that the southern continents are peopled, not by a series of dominant groups corresponding to the Holarctic groups, evolved in a common Antarctic center, but chiefly by groups of more or less tropical affinities and by a few northern groups which have crossed the tropic barrier. There are many groups of birds living to-day in the widely separated tropical regions whose ancestors have not thus far been discovered in the Holarctic Tertiary. But they correspond, both in distribution and in relative position in the classification, with other groups which the geologic record proves to have originated by dispersal from Holarctica, and there is no valid reason for assuming any other origin. The geologic record of Tertiary birds is far more fragmentary than that of Tertiary mammals and especially in the Nearctic region.

It should further be observed that the perching birds represent the primary adaptation from which the various specializations—terrestrial, wading, marine, etc.—have diverged, and that, in consequence, these divergently specialized forms retain various archaic features which have been lost by the central group.

The relations, dispersal and present distribution of birds are thus wholly in accord with the principles here set forth. The detailed application of these principles is beyond the limits of the present discussion.

DISPERSAL OF AMPHIBIA

The modern Amphibia include a few small and for the most part highly specialized survivors of a group whose period of dominance dates back to the Paleozoic. Of their Mesozoic and Tertiary ancestry almost nothing is known. The Stegocephalia, the dominant Amphibia of the Permian, were far less aberrant and much nearer to the contemporary primitive Amphibia; their interrelationships are still far from being precisely definable, and, until these are better understood, it is futile to discuss the evidence which they may furnish as to former geographic connections.

The distribution of the modern Amphibia is often notably discontinuous, and in the absence of evidence from extinct types as to the real origin of these discontinuous distributions they are interpreted by many authors as affording evidence for various transoceanic bridges. But they are not essentially different from various instances of discontinuous distribution among Mammalia, except that they are probably in some cases of more ancient origin, and are less restricted by ocean barriers.

The urodele Amphibia are Holarctic, save for one family, Plethodontidæ, which has spread into northern South America and has also reached Hayti. Although thus limited in dispersal, they would seem to be an ancient group represented as far back as the Wealden by *Hylæobatrachus,* said to be related to the modern *Cryptobranchus.*[116] Their distribution within Holarctica is more or less of a relict type, broken up by the unfavorable environment of so large a part of this region, especially of the central portion. The cæcilians are tropical but have not reached Australasia.

The frogs and toads have a wide dispersal, and so far as a superficial view may show, the most primitive or archaic families are limited to the peripheral continents and oceanic islands, while the more progressive

116. F. Broili, in Zittel's Grundzüge der Palæont., Vertebrata: 176. 1911.

groups are more cosmopolitan, but have not yet reached all of the outlying regions. Some of the families, at least, would appear to be of ancient origin; *Palæobatrachus,* allied according to Gadow[117] to the Aglossa of the Ethiopian and Neotropical regions, is recorded from the Jurassic of Spain, and is said to be common in the older Tertiary of Europe. Among the modern families the Cystignathidæ are chiefly Australasian and Neo-

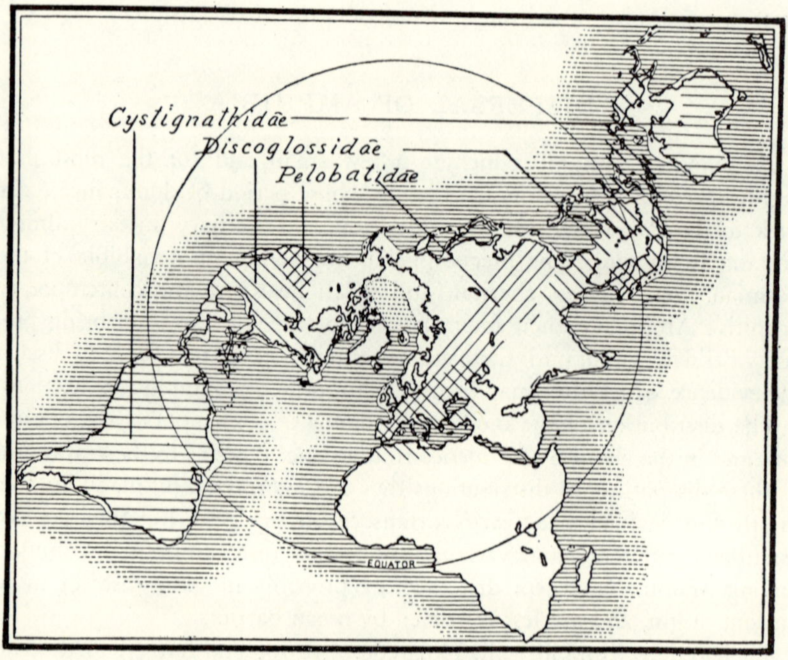

FIGURE 33. Distribution of three families of Anura

These may be interpreted as due to three successive dispersals from the north. The other families of frogs and toads are more widely spread, and their regional abundance has conditioned certain peculiarities in the distributions here shown.

tropical, but a few are still found in North America. This distribution parallels that of the polyprotodont marsupials, except that the latter have not reached New Zealand or the Antilles, or entirely disappeared from the East Indian islands. The Discoglossidæ inhabit the East Indies and North America but have disappeared from the intervening portion of Holarctica; *Discoglossus* and other genera are found in the Middle Tertiary of Germany. The Pelobatidæ stretch across Europe and Asia and northwestern

117. H. GADOW: *l. c.* 145. 1901.

North America. These three families represent evidently three successive dispersals.

The other families are more cosmopolitan. The genus *Bufo* has failed to reach Australasia, Madagascar or New Zealand, but is replaced in Australia by a (more primitive?) member of the family. The Hylidæ are to-day chiefly South American and Australian, but a few members still inhabit North America. They are not found in Africa or the Oriental region, where it seems reasonable to suppose that they have been displaced by the true frogs (Ranidæ), peculiarly varied and abundant in these regions. The Ranidæ, like the Bufonidæ, represent a less ancient dispersal, probably from a southern Palæarctic or Oriental center, since they have reached northern Australia on one side and northwestern South America on the other, and, while they have reached Madagascar and the Solomon Islands, they have failed to reach the Antilles.

These suggested lines of dispersal are based upon the present distribution interpreted in accord with the principles outlined in previous pages of this article. While the past history of the Amphibia is too little known to confirm them by adequate direct evidence, I believe that good inferential evidence might be obtained from a comparison of the progressive or archaic characters of the skeleton in the different families. The fossil Amphibia afford sufficient evidence to determine the broader lines of their evolution and differentiation, although they tell very little about their past distribution. The same conditions hold true with regard to the fresh-water fishes.

DISPERSAL OF FRESH-WATER FISHES

The fresh-water fishes afford many striking illustrations of isolated primitive survivals in the southern continents and especially in their tropical parts. With marine fishes, the distribution is wider, as we should expect, and the dominant types are generally world-wide in their distribution. Yet, even with marine fishes, a superficial survey seems to show the majority of primitive survivals along the southern coasts.

Fishes are, it is to be remembered, dominantly marine. The wider field and more varied opportunities for development afforded by the ocean waters, in contrast with the limited and isolated fields and uncertain tenure afforded by fresh-water rivers and lakes, have conditioned this. The fresh-water habitat for aquatic groups of animals stands in somewhat the same relative position to the marine habitat as does the insular to the continental habitat for land animals. It is the refuge for survivors of primitive faunæ. And, as in the insular land faunæ, we are constantly confronted there with the occurrence in widely remote regions of archaic types apparently nearly related, whose similarity is partly due to independent adaptation to a similar environment, partly to persistent primitivism. *Lepidosiren* in tropical South America, *Protopterus* in tropical Africa, *Ceratodus* in tropical Australia are perhaps the most prominent examples of extremely ancient survivals. These are survivors of early Mesozoic or even Paleozoic marine and estuarine fishes of world-wide distribution, and they have endured, in their tropical refuge, the several successive periods of zonal climate which affected the environment of temperate and tropical regions.

More pertinent to the problem in hand are the relationships of early Tertiary fishes of the northern continents to the modern South American, African (and Australian?) fishes. Here, again, I am compelled to dissent from the interpretations and conclusions of so distinguished an authority as Dr. Eigenmann,[118] who, as it seems to me ignores certain very important parts of the evidence.

There is a marked similarity between certain parts of the fresh-water fish faunæ of South America and of Africa. Eigenmann and others would explain this by a former continental union, but it is certain that some, at least, of these now tropical types existed in the northern continents during the early Tertiary. Eigenmann[119] asserts, indeed, that no part of the modern South American fresh-water fish fauna is derived from North America; but how he reconciles this with the recorded presence of several of the most typical genera in the Green River Eocene of Wyoming, I do not see.

A few cases in point may be noted, as follows:

Lepidosteus, now Central American and southern Sonoran. Abundant in all the Eocene formations of the northwestern States, as also in Europe.

Phractocephalus, Arius, etc., now South American, nearly related to *Rhineastes* of the Bridger and *Amyzon* beds of the western States.

Osteoglossus of Brazil, Borneo and New Zealand, *Vastres* and *Heterotis,* also southern types, closely related to *Dapedoglossus* of the Green River shales (Eocene).

The characins, which form so important an element of the modern South American fauna, are, as Eigenmann holds, largely a local expansive radiation conditioned by the immense ramifying river-systems of that continent. But, considered in their more general relations, they are a primi-

---

118. See especially C. H. Eigenmann: "Fresh-water Fishes of Patagonia," Reports Princ. Univ. Exped. Patagonia **3**(3-4): 1909-10.

119. C. H. Eigenmann: Popular Science Monthly **1906**: 523.

tive group, the northern cyprinids being a higher and later development.

The catfish, which in the North have the characteristics of a disappearing group, are numerous and dominant in South America. Eigenmann calls attention to the paucity of the Patagonian fauna and its apparent relations to that of New Zealand and Australia (Galaxiidæ and Aplochitonidæ). He does not, however, attach any great weight to this as evidence for a former Antarctic connection, regarding it as "highly theoretical and precarious" so far as the fresh-water fish are concerned—but "The evidence from other sources of a former land connection has become conclusive." I might observe here that many students in other groups are equally doubtful of the conclusiveness of the evidence for Antarctic connections in the groups with which they are familiar, while equally ready to accept as conclusive the evidence in groups with which they are not familiar.

As regards a connection of tropical Africa with tropical South America, Eigenmann is much more positive, basing it mainly upon the characins and cichlids, common to both continents. There is no species or genus common to the two continents. Both families are relatively primitive, as compared with northern related groups. As regards their former presence in the northern world (which Eigenmann does not allude to) or their parallel adaptation from marine forms of Cretaceous or early Tertiary time, there is little satisfactory evidence. Nevertheless, the fact that they represent an adaptive divergence from an intermediate and more primitive type ancestral to carp and catfish is a suggestive one.

If now we compare the general relations of tropical fresh-water fishes with those of the North, it will appear very clearly that the highest and latest in appearance of the several groups are still limited to the northern world, and that, in the tropics, more primitive groups exist, many of them known to be former residents of the northern world, others much nearer to known or inferred ancestral groups than are any members of the present northern fish fauna. Where the environment favors, some of these groups have branched out into an immense variety and number, far exceeding what is known in the colder North. But they are distinctly less progressive. In the southern continents, we meet with some remarkable parallelisms to the dominant types of the North, very suggestive at first of Antarctic connections, but probably explainable (as in *Galaxias*) in other ways. These groups impress one as highly progressive, although less so than the northern groups; but they do not appear to have contributed materially to the tropical faunæ.

In some respects the fresh-water fishes present nearer analogies to the birds than to mammals in their distribution; and this is no doubt conditioned by their less strict limitation to land connections for their migration, and to the greater antiquity of the class.

GENERAL CONSIDERATIONS ON THE DISTRIBUTION OF INVERTEBRATES AND PLANTS

It would be unwise to attempt any survey of the paleogeographic data afforded by invertebrates and plants. Lacking both the special knowledge necessary for a critical consideration of the data, and the time necessary to make even an adequate compilation, it would add nothing to the argument. While, for reasons already given (page 103), placing most weight on the evidence obtainable from mammals, I fully recognize the importance and variety of evidence outside the Vertebrata, and the force which attaches to cumulative evidence from several independent sources. At the same time I must express a strong conviction that the sources are not really independent, and that concordant results in several groups which flatly contradict the results obtained by a study of mammals, can only indicate one of two things. Either the interpretation of the evidence among the Vertebrata is incorrect or there are factors of error common to the interpretation of the several other groups which accord in their disagreement. What these factors may be, I have already indicated and have attempted to show that they account for discordant results based upon the distribution of the lower vertebrates and interpreted as involving radical changes between continental and abyssal regions which are highly improbable, to say the least, from a geological point of view, and which are not merely unnecessary but apparently impossible when we attempt to explain the distribution of the higher vertebrates in accordance with them.

It is true that the evidence against such changes in pre-Tertiary times is less weighty, and that it diminishes further in the older periods of geologic time. And the antiquity of many groups of invertebrates, especially of land invertebrates, makes it impossible to limit the hypothetical land bridges which their distribution is supposed to require, to the Tertiary or even the Mesozoic. The permanency of the ocean basins in the older geologic epochs is beyond the limits of this discussion.

So far as a superficial acquaintance shows, the general distributional relations of most land Invertebrata and of plants appear to me to accord with those of the Mammalia. Primitive and archaic[120] types abound chiefly in the tropics. The most progressive and dominant types are Holarctic. The southern continents show common groups suggestive of an Antarctic radiation, but which may, like the marsupials or the chrysochloroid insectivores, be remnants of formerly cosmopolitan groups whose resemblance is due rather to persistence or to parallel evolution under similar climatic stimulus than to such close affinity as would involve Antarctic continental connections.

Where, as in the earthworms, we have no knowledge at all of their past distribution, it is impossible to test this interpretation of their present distribution; nor in such a group does it seem possible to estimate how much and in what manner slow progressive climatic change might affect their structural evolution, although climatic conditions are evidently important in controlling their range.

The point that I desire to emphasize is that, if such an interpretation as I have suggested be possible, it should be accepted in preference to one which would involve such unexplainable difficulties in the distribution of the higher animals and such improbable physiographic changes. No hypothesis can be finally accepted that does not conform to the facts of distribution in all groups of animals and plants. It is not a matter of preponderant evidence. Every anomaly must be explained, every distributional fact must be interpreted in accord with the rest, before we can consider theories of paleogeography as conclusively proven. It is not sufficient that the evidence in one group or in ten groups has been interpreted on concordant lines, so long as there remains an eleventh group which cannot be so interpreted. But, pending a final agreement in our deductions from the evidence afforded by the various classes, it appears to me that we should hold to conservative views rather than adopt hypotheses of continental relations so much at variance with generally accepted geological principles and inferences.

To illustrate the point that these discrepancies are a matter rather of interpretation than data I may venture to discuss one or two instances among invertebrates prominently used in paleogeography.

Interpretation of Distribution Data of Crayfish

I am indebted for my data on this interesting group to Dr. Ortmann's valuable discussion of the geographical distribution of fresh-water De-

120. Archaic is used in the sense of divergently specialized but little progressive.

capoda.[121] The interpretation, however, which I would place upon the facts differs widely from his.

As Professor Huxley has observed, the real difficulty in explaining the distribution of the crayfish is in their occurrence in the north and south temperate zones, separated by a wide tropical belt in which none now occur or are known to have occurred in the past. Two explanations offer themselves:

(1) Independent adaptation from marine types in the northern and southern hemispheres. This would involve either former Antarctic connections or independent adaptation also of the several southern groups from marine types.

(2) Former cosmopolitan distribution of crayfish, with subsequent disappearance from the tropical belt and differentiation of the isolated southern groups and of the more progressive northern groups.

The latter view is generally accepted, and seems to me more consonant with the facts of distribution, *e. g.*, presence of crayfish in Madagascar, while they are absent from South Africa. I am unable to agree with Dr. Ortmann that crayfish on oceanic islands necessarily involve a former land connection, since such land connections as he finds it necessary to postulate would apparently involve the presence on these islands of continental faunæ which are not now present, and whose absence cannot be reasonably accounted for. For the reasons already presented I see no difficulty in supposing that the crayfish of Cuba, Madagascar, New Zealand or Fiji have reached those islands by accidental transport of natural "rafts" through the agency of ocean currents, or by other accidental means. The Australian and South American crayfish I should regard as derived from the north, by way of the existing or slightly submerged land bridges, at a time when the northern crayfish were much more primitive than now, and when, for reasons which I do not venture to suggest, the tropics were a more favorable environment than now. The northern crayfish have since evolved into *Potamobius* and *Cambarus,* the southern specialized into the more divergent *Parastacus* of South America, *Cheraps* and *Eugæus* and *Astacopsis* of Australia and Tasmania, *Paranephrops* of New Zealand and ? Fiji and *Astacoides* of Madagascar.

Of these southern genera, *Astacoides* is the nearest to the northern types. This is to be expected, if the southern genera are remnants of a cosmopolitan distribution derived by dispersal from the north; for the Malagasy genus would be a derivative from Ethiopian crayfish, which would be less remote from the north, and would be correspondingly more advanced

121. A. E. ORTMANN: "Geographical Distribution of Fresh-water Decapods and its Bearing upon Ancient Geography," Proc. Amer. Phil. Soc. 41: 267-400. 1902.

than in South America or Australia. As far as the more special distribution of the northern crayfish is concerned, Dr. Ortmann's paper affords data for the following interpretations.

Two genera are concerned, *Cambarus* of the eastern Sonoran region, and *Potamobius* (*Astacus* of most authors) of the Old World and western Sonoran region.

In his discussion of the genus *Cambarus* Ortmann states that the more primitive forms of the first, second and fifth groups belong chiefly to the south towards Mexico, and interprets this as meaning that the genus came *from* Mexico. But, according to the principles here adopted, this should mean that the center of dispersal is to the north and east; and the discontinuity in range to the south and west is exactly what we should expect, Dr. Ortmann's attempt to find an explanation for it on the opposite theory of migration being curiously complex and unconvincing. The most primitive species occur in such widely divergent points as Mexico and Cuba.

The more primitive genus *Potamobius* has a more discontinuous range, in Europe, part of Eastern Asia and Western North America, the Asiatic species being nearest to *Cambarus* (*i. e.,* highest in development) but parallel, not truly closely related. This, I take it, is correctly interpreted by Ortmann as indicating an Asiatic center of dispersal for this genus. But in place of supposing with Ortmann that *Cambarus* originated from species of *Potamobius* pushing down southward into Mexico and thence northward again (as *Cambarus*) into the United States, it seems to me that the rational explanation would be to suppose that both genera are the disconnected remnants of a formerly Arctic center of dispersal. This would be first split in two by a progressively unfavorable environment, one division passing down into America east of the Cordilleras, and developing into *Cambarus,* the other part in Asia progressing more slowly into *Potamobius* and spreading east and west from that center, as the American group spread southward.

Distribution of Helix hortensis

Dr. Scharff[122] regards the distribution of *Helix hortensis* as an important part of the evidence in favor of a late Cenozoic bridge connecting Europe with eastern North America. The species is well known in Europe and has always been regarded as indigenous there. It occurs along

122. R. F. Scharff: Proc. Roy. Irish Acad. 28: 19. 1909.

the North Atlantic coast, and in Labrador, Greenland, Iceland and the Shetland and Faroe Islands. It was formerly considered as introduced on this side of the Atlantic by human agency; but it has been found in old Indian shell-mounds and more recently in undoubtedly Pleistocene deposits in Maine. It is unknown in Asia or western North America. Hence, Dr. Scharff concludes that it must have migrated from Europe to America across a land bridge via Iceland and Greenland in Pliocene or Pleistocene times.

The early opinion that *Helix hortensis* is an introduced species in this country was founded, so far as I recall, mainly upon the peculiar local range and habitat of the species, very different from the truly indigenous New England land-snails, and my early experiences in land-snail collecting in southern New Brunswick were quite in accord with this evidence. It is quite possible that *Helix hortensis,* like the genus *Equus,* is both introduced and indigenous.

Granting that it is at least partially indigenous, what evidence is there that the present distribution is not the remnant of a Tertiary circumpolar distribution? The fact that it is not recorded in the Tertiary of Asia? But what proportion of the presumably abundant Miocene or Pliocene land-snails of Asia is known to us? It can only be a minute fraction at the best—less than one per cent. So the chances are a hundred to one that if *Helix hortensis* or an ancestral form of the species existed in the Tertiary of North Asia, we should have no record of its existence at present. We do, however, have a good deal of indirect evidence that an environment favorable to the present habits of the species existed during the later Tertiary in the region intervening between its present discontinuous distribution areas, and that the environment became unfavorable in that intervening region at the close of the Tertiary. I can see no need for assuming a transatlantic land bridge to account for the distribution of this species. And the explanation here suggested is in harmony with the known course of distribution of those members of the northern land faunæ whose past history is preserved to us in the geologic record. It involves only those minor changes of continental level (a few hundred feet) of whose occurrence during the Pleistocene we have ample evidence.

On the other hand, if we assume such a transatlantic land bridge during the late Tertiary we must suppose an elevation of upwards of 5,000 feet, a huge disturbance of the isostatic balance of whose possibility we have no real evidence; for the submerged channels so often cited in support of these immense uplifts have been shown by Chamberlin to be much more probably due to "continental creep," to the slipping down, so to speak, of marginal sediments to a lower level.[123] In any case, there could be no evidence as to the period at which these old channels were last above water. They may have been submerged since the Permian, for aught we know to the contrary. Furthermore, we have to explain the non-migration of a multitude of forms which got just so far as conservative land elevations could carry them, but no farther.

Distribution of Percidæ

Another instance upon which Dr. Scharff lays great stress is the distribution of the perches. Here, the false impression produced by the use of a Mercator's projection map in plotting the distribution of northern forms, seems to me to be very obvious. This map does not give the northern regions in their true proportions or relations. Transferring the distribution of this family as plotted by Tate Regan, to a north polar projection map we get the real relations and proportions with approximate correctness. It then becomes obvious that the perches are centered around the drainage basin of the Arctic Ocean. In North America they have extended down the Atlantic coast drainage area and into that of the Gulf of Mexico as far as the Rio Grande. In Asia they have been admitted by the old Hyrcanean Sea into the present Caspian and Aral basins; and a glance at the late Tertiary geography of Europe will show how they have reached the drainage basin of the northern Mediterranean. They are not now found in the Arctic drainage area of western North America, Greenland or Iceland, where the environment, now or in the Pleistocene, is amply sufficient to account for their extinction. What need of a transatlantic land bridge to account for this distribution.

A fourth instance cited by Dr. Scharff is the distribution of the river-mussel *Margaritana,* and as he well observes, numerous other instances would probably show similar discontinuous distribution. But, so far as I have been able to find such instances, the same reasoning and the same explanation apply to them all.

CRITICISM OF SOME OPPOSING HYPOTHESES

It is not practicable to take account here of the flood of paleogeographic

123. There is another possible explanation. The progressive building out seaward of barrier reefs around a number of separate centers until they joined into a platform would naturally leave deep intervening channels, especially off the mouths of great rivers where the influx of mud and fresh water hindered the growth of the coral organisms. The submarine contours around the West Indian islands especially suggest this explanation, which I offer tentatively for the consideration of my better-versed confrères.

discussions of recent years which have advocated all sorts of consistent or inconsistent changes in continental outlines. They agree for the most part in failing to take into account certain considerations which to my mind are essential elements in any problem of distribution.

Among the geological considerations are the following:

(1) Evidence that the present distribution of the deep ocean basins is in the main due to isostatic balance. This affords a strong presumption in favor of its permanence.

(2) Absence of abyssal deposits in the geological formations of any continental region. Chalk deposits are not an exception, as it has been shown that they were deposited in shallow epicontinental seas rather than in deep oceanic basins.

(3) Abrupt ending of an elevated line of disturbance and its continuation as a submerged line of disturbance does not necessarily indicate that the submerged portion was formerly elevated, although it does reduce the improbability of its former elevation by indicating a line of disturbance and hence of possible elevation.

(4) The presence of marine formations of Cretaceous or Tertiary age over large portions of the interior of the great continents does not indicate that these continents first came into existence as such during the Cretaceous or Tertiary. In the better-known portions of the earth's surface we know well enough that these marine formations were due to periodic temporary submergence, interrupted by periods of more or less complete emergence. It is but reasonable to apply the same explanation to the less known regions. I see no more reason to suppose, as do Von Ihering, Scharff and others, that South America first came into existence as a united continent in the Tertiary, than to conclude on similar evidence that North America was but a group of isolated land masses until the end of the Cretaceous. In this country, we have positive proof of its antiquity; but the evidence for recent origin of the South American would apply just as well to the North American continent. A similar presumption of antiquity applies to Australia, Asia and Africa.

Among zoölogical considerations we may mention the following:

(1) The discontinuous distribution of modern species is again and again taken as proof that the regions now inhabited must have been connected across deep oceanic basins, without considering the possibility that it is a remnant of a wider past distribution, or that it is due to parallel evolution from a more primitive type of intermediate distribution, now extinct. Yet so many instances are known where the geological record has furnished proof that one or other of these explanations applies to cases of discontinuous distribution, that it would seem that these ought to be the first solutions of the problem to be considered, and that in view of the known imperfection of the geologic record, mere negative evidence is not sufficient to cause them to be set aside.

(2) No account is taken of faunal interchanges often much more extensive, which would presumably have taken place if the land bridges assumed had existed, but which have not taken place. It may here be urged that this too is negative evidence. But the negative evidence derived from an appeal to the geological record is weak, not *per se,* but because of the demonstrated imperfection of this record. On the other hand, there are many instances where a land bridge is well proven, and in these cases it is not a few scattered exceptions but an entire fauna that has migrated, subject only to the restrictions imposed by climatic or topographic barriers of other kinds.

I may venture upon a discussion of a few instances in order to show the type of objections which appear to me to apply to much of the evidence cited in favor of most of these transoceanic land bridges.

ON VAIN SPECULATIONS

According to some distinguished paleontologists,[124] progress is to be made only by ignoring the possibility that races have originated in or migrated from regions of whose former life we have substantially no record, and assuming that they must have evolved in one or another region where the record is more or less known, and that the actual record must be the sole basis for any conclusions. They refuse to consider the arguments for origin elsewhere, on the ground that such hypotheses are "vain speculations" and "serve merely to conceal our ignorance."

To this I may answer that a fair and full consideration of the data at hand shows that such hypotheses, of one kind or another, are absolutely necessary, unless we are to abandon all belief in the actuality of evolution and are to treat it as merely a convenient arrangement of successive species and faunas independently created. Such a view was held by Agassiz and most of his predecessors, but it is unnecessary to consider it in the present state of scientific belief.

If, on the other hand, we accept the belief that the successive species of each phylum are genetically related, how are we to explain the fact that these phyla are usually approximate and not direct, and that where the evidence is most complete, the fact that they are not in a direct line of structural evolution stands out most clearly. Take for example the ancestry of the horse, the most striking, easily recognizable, widely known and thoroughly studied illustration of mammalian evolution. It was possible, when the "documents" were few and imperfect, to trace a supposedly direct line of ancestry through European predecessors. Later, when the fossil fields of the western United States were first explored, a much more direct line of ancestry was found in this country, and the European series was recognized as not being the direct line. But the further progress of exploration in America, and the discovery of complete skeletons of the supposed ancestral stages known at first only from fragmentary specimens, has demonstrated that this line too is an indirect and approximate series so far as the succession of the known species is concerned. This has been recognized in recent years by American students, and variously phrased or interpreted. The most probable explanation of the facts is to suppose that the known phylum is approximate, not direct; that the direct line of descent leads through unknown or imperfectly known species, and that those known to us are offshoots of varying closeness. The direct line is,

---

124. Depéret, Thévenin and others.

then, admittedly through hypothetical species, and the only question is whether the habitat of these species was in the regions where we have searched vainly for their remains, or in the much greater intervening region where we have not searched. Horses are found throughout the Tertiary in central and western Europe on the one hand, on the Western Plains of America on the other. There is every reason to believe that they inhabited all or parts of the intervening region and we have no right at all, in weighing the evidence, to refuse to take this region into consideration, on the plea that it has furnished no "documents" as yet. To place such limitations on our theories would hardly tend to solving our problems, however much it might seem to simplify them. It is merely to prefer a conclusion that we know to be false to a conclusion that we cannot prove by direct evidence to be true.

What I have stated in regard to the fossil ancestry of the horse applies to most mammalian phyla, in greater or less degree according to the perfection and number of our "documents." Where these are few and fragmentary, it is still possible to build up phyla which cannot be proven to be inexact. But, as knowledge increases and becomes more exact, these phyla are more and more broken and complex, and direct genetic series become more limited in extent. This is to be expected, for the regions which up to date have been at all thoroughly explored are but a small fraction of the area which the group concerned must have inhabited. And on *a priori* grounds, the chances are greatly against the particular species which was to become dominant inhabiting the particular regions which we have explored.

Professor H. F. Osborn has very well expressed the conditions of evolutionary progress by stating that each group is highly polyphyletic, consisting of numerous subphyla evolving along more or less parallel lines. But we are here concerned less with the disentanglement of the subphyla of a group than with its dominant center of dispersal as a whole. And from this point of view it seems to me misleading and erroneous to assume that it must have migrated only from one to others of the regions where its remains have actually been found, instead of attempting to locate from the indirect evidence available the true center of dispersal.

In contrast with the views here criticized, I may venture to quote from an address in which Dr. Stehlin[125] has recently summarized the phylogenetic results of his monumental studies upon the Eocene fauna of Egerkingen, a work of extraordinary thoroughness and ability which, as a

125. H. G. STEHLIN: Verh. Schw. Naturf. Gesell. **93** Jahresversammlung, Sept. 1910: 29 of separata.

recent reviewer observed, has involved a revision of the entire Eocene mammal fauna of Europe: "Where then dwelt these yet unknown herds of mammals evolved during the Eocene, whose existence is recorded through their influence upon Europe and North America the more clearly as we analyze more closely the data obtained in these continents? We can scarcely be wrong if we look to the huge continental mass of Asia, still almost unexplored by the paleontologist. The future, and, it may be hoped, the near future will show how far our present anticipations are correct."

SUMMARY OF EVIDENCE

The geologic evidence for the general permanency of the abyssal oceans is overwhelmingly strong. The continental and oceanic areas are now maintained at their different levels chiefly through isostatic balance, and it is difficult to believe that they could formerly have been reversed to any extensive degree. The floor of the ocean differs notably in its relief from the surfaces of the continents, and only in a few limited areas is the relief suggestive of former elevation above sea-level. The continental shelf is so marked, obvious and universal a feature of the earth's surface that it affords the strongest kind of evidence of the antiquity of the ocean basins and the limits beyond which the continents have not extended. The supposed evidence for greater elevation in the erosion channels across its margin have been shown to be better interpreted as due to "continental creep." The marine formations now found in continental areas have all been deposited in shallow seas. No abyssal deposits have ever been certainly recognized among the geologic formations of the continental platform.

Leaving out of consideration speculative hypotheses as to a formerly smaller amount of water on the surface of the globe, shallower ocean basins in Paleozoic times and different land and water distribution in the older geological periods, it is sufficient for the purposes of this discussion to emphasize the great weight of geological and physiographic evidence for the permanency of the continental masses as outlined by the continental shelf, during the later geological periods, and especially during the Tertiary.

The present distribution of continents and oceans on the surface of the globe (as outlined by the continental shelf) consists of a great irregular northern mass including Europe, Asia and North America, with three great partly isolated projections into equatorial and southern latitudes,

South America, Africa and Australasia, and a smaller Antarctic land mass wholly isolated. The three peripheral continents are isolated from each other and from the Antarctic land by broad and deep oceans, but with the doubtful exception of Australasia, are united to the central mass by shallow water or restricted land connections.

A rise of 100 fathoms would unite all the continents and continental islands, except perhaps Australia, into a single mass, but would leave Antarctica, New Zealand, Madagascar, Cuba and many smaller islands separate. A further elevation of five times this amount would not alter materially the boundaries of land and sea. A submergence of 100 fathoms would isolate the three southern continents, and cause shallow seas to spread widely over the interior of all the continental masses, reducing some of them to isolated fragments or archipelagoes.

Such cyclic alternations of emergence and overflow are recognized by many geologists as the dominant feature of the earth's history, corresponding to the succession of periods into which geologic time is divided. The greater disturbances resulting in folding, faulting and mountain making, while involving much greater changes of level, affect more limited areas, adjacent to lines of unstable equilibrium, especially along the borders of the continental platforms.

Associated with these great cycles of elevation and submergence are climatic cycles from extremes of cold or arid zonal climates culminating in glacial epochs, to the extremes of warm humid uniform climates which accompany or follow the extremes of submergence.

The effect upon terrestrial life of progressive elevation of the land areas, accompanied by a progressively cold climate at the poles and arid climate in the interior of continents, would be to adapt the terrestrial life to cold, arid and highly variable climatic conditions. The environment favorable to this adaptation will appear first near the poles, and the northern and southern faunæ will be more progressive and will tend to disperse towards the equatorial regions. The wider area of emerged continents will tend to expansive evolution of the land faunæ, and their union into a single land mass will facilitate cosmopolitan distribution. Owing to the conformation of the continents the dispersal will be chiefly from the Holarctic region, the Antarctic and southern lands being unfavorably situated for the evolution and dispersal of dominant races and contributing but little to the cosmopolitan faunæ of the emergent phase. These conditions are also favorable to the development of higher, more active and more adaptable types of terrestrial life, which tend to supplant even in moist tropical regions the less adaptable remnants of the tropical faunæ which find there their last refuge.

During the opposite phase of the cycle, the faunæ become progressively readapted to the moist tropical climatic environment. But owing to the higher evolutionary stage acquired during the arid phase, the higher and dominant types of the new fauna are evolved chiefly by readaptation of the dominant types of the arid phase and only subordinately by expansive evolution of the tropical fauna surviving through that phase.

The paleontologic record appears to be in exact harmony with these principles, provided due allowance be made for its imperfections. The geographic distribution of animals and plants affords far more complete data, but their true significance has in my opinion been misinterpreted by many zoögeographers. When interpreted in harmony with the principles of dispersal shown to be true among mammals, they yield fully concordant results. The geologic record is to-day far more incomplete than is generally admitted, and will always be incomplete. Negative evidence, while sometimes of high value, is more often worthless and should never be admitted without a careful canvass of the situation in each instance.

The population of oceanic islands is notably incomplete and cannot be interpreted as due to continental connection. The difficulties in the way of over-sea transportation are best explained by the hypothesis of natural rafts; the degree of probability that attaches to this hypothesis is estimated.

The dispersal of mammals is then considered at some length, order by order, and it is shown to accord fully and in detail with the principles here set forth, and to be impossible of explanation except upon the theory of permanence of the ocean basins during the Cenozoic Era. While the prominence of the Holarctic region as a center of dispersal is ascribed to its central position and greater area, some evidence is given to show that climate is also a factor in the greater progressiveness of the northern, since it is also noticeable in the southern as compared with tropical faunæ.

The distribution of the Reptilia appears to be in conformity with the principles here outlined, and extends their application to the Mesozoic Era. The distribution of birds and fishes and of invertebrates and plants is probably in accord with the same general principles, modified by differences in methods of dispersal. The opposing conclusions that have been drawn from the distribution of these groups are believed to be due to an incorrect interpretation of the evidence. A few instances, which have been prominently used to support opposing conclusions, are analyzed and shown to conform to the conclusions above set forth, if interpreted upon similar lines as the data for mammalian distribution.

APPENDIX

Since this paper was written two very readable and instructive books on geographic distribution have appeared, "The Wanderings of Animals" by Professor Gadow,[126] and "Distribution and Origin of Life in America" by Professor Scharff.[127] Both writers, and especially Doctor Scharff, belong to what may be called the bridge-building school of paleogeography, and the general criticisms expressed in the earlier part of this article apply largely to their interpretations. It is with no intent to depreciate their value that I observe that there are numerous errors of fact in those portions of the evidence with which I am best acquainted, for in a subject of so wide a scope most of the evidence is necessarily compiled and not very well understood, and errors more or less essential will slip in. It is for that reason that I have avoided detailed discussion of the parts of the evidence on the present subject with which I am not well acquainted; and, in spite of a good deal of checking and revision, I have no doubt that the foregoing discussion contains various inaccuracies.

A more serious criticism is the illegitimate and often partisan use made of negative evidence. This is doubtless due to the same cause, a mere book knowledge of the fossil record, and failure to examine and weigh its evidence. But it is very obviously affected by a readiness to rely on negative evidence that favors their theories and to ignore a vastly greater amount of negative evidence that does not.

Dr. Gadow considers it "awkward" for the theory of Holarctic dispersal of the marsupials in the Cretaceous that no survivors have been recorded in the Tertiary of Asia. He prefers to believe that the Australian marsupials arrived via Antarctica from South America. If it is "awkward" for the one theory, that although survivors are found in the early Tertiary of both Europe and North America, none have been found in Asia, then it must be equally "awkward" for the theory that Dr. Gadow supports that none have been found in Antarctica. For we know even less about the early Tertiary of Asia than we do about the Antarctic Tertiary. If the absence of zalambdodont insectivores in the Eocene of Europe is to be

126. HANS GADOW: "The Wanderings of Animals." Cambridge Manuals of Science and Literature 64: 1913.

127. R. F. SCHARFF: Distribution and Origin of Life in America. Macmillan Co., publishers, New York. 1912.

assigned any weight, then equal weight should be assigned to their absence from the Oligocene and Eocene of South America and from the Pleistocene of Cuba, of Madagascar and South Africa. We know as much about the one fauna as we do about the others. The negative evidence has no weight in any of these instances; *per contra,* the fact that zalambdodonts are known to have lived in the early Tertiary of North America (Paleocene to Oligocene) affords a presumption of their presence in the nearly allied early Tertiary faunas of Europe, just as their presence in the recent faunas of Madagascar and South Africa and in the Miocene of South America affords a presumption of their presence in the nearly allied faunas which immediately preceded them. Equally, the presence of marsupials in the early Tertiary of Europe on one side of Holarctica and of North America on the other side raises a strong presumption of their presence in the intervening region of Asia from which no fossils are known. They are not found in the later Tertiary of Europe and America, so that we should not expect to find them in the later Tertiary of Asia. On the contrary, the small fragment of evidence that we have as to the Tertiary fauna of Antarctica affords a slight presumption against the presence of mammals on that continent.

Doctor Gadow's statement that the Chiroptera did not reach America until the Pleistocene is another curious instance of the misuse of the fossil record, which no one familiar with the character of our Tertiary formations and the necessary limits of the fossil faunas would be likely to make; nor would anyone acquainted with the variety and specialization of the New World genera be inclined to believe that it was all the result of post-Pliocene immigration and differentiation. Most of the creodonts, he informs us, "died out with the Eocene or rather they were modernized into the typical Carnivora in various parts of the world. Some, however, kept on to almost recent times as highly specialized creodonts, *e. g.* the sabre-toothed tigers: *Nimravus* in North American Oligocene; *Machærodus* from Miocene to Pleistocene in Europe and Asia, whence in the Pleistocene it appeared as *Smilodon* in America. . . ."[128] It is perhaps unnecessary to point out that the machærodonts were not creodonts but typical Carnivora of the family Felidæ, and that their evolutionary series is fully as complete and progressive in the Nearctic as in the Palæarctic record. I may also note that "small swine" (meaning I suppose the primitive bunodont artiodactyls from which both pigs and peccaries are derived) appeared in North America quite as early as in Europe; that the genera

128. *Op. cit.*

Procamelus and *Pliauchenia* do not mark the splitting of the Camelidæ into camels proper and llamas; that *Dorcatherium* is not identical with *"Hyomoschus"* (*Hyæmoschus*) and is an older name; that *Arsinoitherium* is not a pair-horned dicerathere but is a representative of a distinct order of mammals; that the precise relations of the American Eocene tapirs have yet to be determined; that *Protapirus* does not first appear in the Lower Oligocene of Europe but in the Mid-Oligocene of Europe and North America; that there is no reason to believe that the European *Paratapirus* is more directly in line of descent of the later tapirs than is the so-called *Tapiravus* of the American Miocene, and that the very fragmentary and inadequately studied record of the evolution of the Tapiridæ is quite inadequate for the positive and exact statements which Gadow makes as to their "wanderings."

The statements as to the evolution of the horse show a surprising amount of inaccuracy, considering that this is so widely known a story. Apparently, it is in part the result of an attempt to criticize and modify the conclusions of American writers on the basis of a hasty survey of the incomplete materials available in European museums. The Eocene ancestors are disregarded, because they "are still so very generalized that they lead to horses, rhinos and tapirs as well as to other distinct groups." While this is not far from the fact as regards the Lower Eocene *Eohippus,* it certainly is not true of *Orohippus* and *Epihippus* of the Middle and Upper Eocene. The relations of *Miohippus* to *Mesohippus* are hardly to be dismissed with a "perhaps." *Desmatippus* is not an ancestor of *Parahippus* but is identical; *Hypohippus* is not intermediate between *Para-* and *Merychippus* but is an aberrant type descended from *Miohippus* through *Anchitherium;* the American Miocene series does not come to an end with *Merychippus,* but this genus gives rise through numerous intermediate species to *Protohippus, Pliohippus* and *Hipparion. Hippidion* is not a descendant of *Hipparion* but of *Pliohippus.* There is, it is true, a considerable gap between *Hipparion gracile* and *Equus,* this species being too specialized in tooth pattern and its lateral digits exceptionally heavy; but most of the American hipparions are simpler and less aberrant in tooth pattern and the shafts of their lateral digits reduced often to mere threads. The proximal splints in these forms are very nearly as much reduced as they are in *Equus;* the gap which Doctor Gadow declares has been "slurred over" lies simply in the fact that no specimens have yet been found in which the shafts of the lateral digits are discontinuous but the distal rudiments preserved. Anyone familiar with the difficulty of securing proof of this condition in a fossil species, and with the imperfection of our record of the Pliocene Equidæ, will hardly consider this as a serious gap. Certainly, it is trifling in comparison with the gaps in any of the other mammalian phyla which Doctor Gadow accepts without difficulty. As for the derivation of *Equus* from primitive species of *Hipparion* rather than of *Protohippus,* my opinion to that effect rests upon intensive studies of Miocene Equidæ undertaken for Professor Osborn's monograph of the Evolution of the Horse (in preparation) and I do not think it fitting to publish the evidence in its support at present.

The sirenians, Dr. Gadow tells us, afford strong support of the theory of a transatlantic bridge, the earliest being known from the Eocene of Jamaica and Egypt, etc. They would, undoubtedly, if there were sufficient reason to believe that they were absent from the more northerly parts of the North Atlantic-Arctic shores during the early Tertiary. But there is none whatsoever; the North Atlantic coasts either extended during the Tertiary beyond their present limits to or towards the continental shelf, or else their marine and littoral deposits have been destroyed by glaciation; at all events none remain above water worth mentioning from New Jersey on one side around to the British Isles on the other. That no littoral vertebrates should be known where there are no littoral deposits is not surprising; yet it is upon this worthless negative evidence that the "strong support" rests.

I have limited myself in the foregoing criticism to noting a few points in regard to fossil mammals. Dr. Scharff's book is far too extensive for any detailed criticism here, even within these limits. I can note only that, while highly instructive as well as entertaining, it is far from being either accurate or fair in its treatment of the geological aspects of the subject or the fossil record. The view that during the Glacial Epoch the glaciers were confined in this country to the higher mountain ranges[129] is one that even a biologist is hardly excusable for upholding. Nor does it seem that anyone discussing the Tertiary geography of North America should be so little informed as to suppose[130] that the eastern and western portions of the continent were separated during the Eocene by an ocean barrier. In his argument against the permanency of the ocean basins, Scharff is, on the other hand, able to quote high authority. But the weakness of the argument is nevertheless apparent. That there have been great changes of level along certain lines of disturbance has never been questioned. But the conclusion that the continental platforms have never been submerged to abyssal depths, based upon the entire lack of abyssal deposits in their

129. *Op. cit.*, 46 ff.

130. *Ibid.*, 357.

geological succession, is not disproved but rather confirmed by the recognition of abyssal deposits on an oceanic island lying along a line of high disturbance. For that merely proves that abyssal deposits are recognizable as such when they do occur, absence from the continental platforms remains untouched. Nor does the occurrence of ancient sedimentary and metamorphic rocks on some, especially of the larger, oceanic islands afford any evidence that they are remnants of former continents. The same processes of sedimentation, regional metamorphism and orogenic upheaval must of necessity occur in any oceanic island of considerable size and antiquity, and produce similar results both stratigraphic and petrographic. Moreover, if such islands lie in a line of disturbance which is continued under the ocean to an adjacent continent the same earth-movements may well affect both areas without raising the intervening region above the abyssal depths in which it now lies.

Dr. Scharff adopts Ameghino's correlations of Argentine formations, and von Ihering's assertion that the continent of South America did not exist as a single land mass until late in the Tertiary. I may note by the way that Ortmann[131] not long ago, in reviewing Pfeffer's[132] essay on the zoögeographical relations of South America, rebuked him severely for not being aware of this "undoubted fact," which he declared was not a theory at all. The real facts are that marine and fresh-water formations of Jurassic, Cretaceous and early Tertiary age occur extensively in the interior of South America, indicating that the broad low-lying interior of that continent was periodically flooded by shallow seas. The conditions parallel those of the North American continent very closely, so far as they are known. The North American continent we know existed as such throughout geological time, although extensively flooded at times by shallow seas, especially during the Middle Cretaceous. The same is presumably true of South America.

Like Doctor Gadow, Doctor Scharff makes a wholly unjustifiable use of negative evidence where it may serve to support his views. He is a much more reckless bridge-builder, and appears to be quite unconscious of any difference in probability between such a bridge as the Alaska-Siberia connection and the various trans-Atlantic and trans-Pacific bridges which he invokes. Yet the Alaskan bridge is in existence to-day, only a few yards of its planking removed, if one may so speak, the substructure intact, and the marks of the missing planks still showing on the undamaged portion, while the huge bridges which he "prefers" to believe in are,

131. A. E. Ortmann: Amer. Nat. 39: 413-416. 1905.
132. G. Pfeffer: Zoöl. Jahrb., Suppl. 8: 407-442. 1905.

except for the Icelandic ridge, scarcely indicated by so much as a sandbank on the flat abyssal floor of the vast intervening oceans. That he can claim support of a kind from so high an authority as Suess may be true, but scientific problems should be settled by examination of the evidence, not by citations of opinion from selected authorities.

Doctor Scharff does not at all believe in accidental transportation by floating vegetation or other natural means. Why, he demands, do not the advocates of such views cite instances of such transportation in modern times, and why is it only the more ancient animals that are so transported? The argument is curiously parallel to the favorite anti-evolutionist demand. Why, if man has evolved from a monkey, do not the scientists take a monkey and turn him into a man? Of course, the proof demanded is an impossibility. If any instances of such transportation were noted during the last few centuries, they would be ascribed to human agency; but the probabilities within that time are slight except in islands near the coast, such as Krakatau; for more distant islands they are made probable only by the vast length of geological periods, and it is a matter of course that the more ancient the type, the longer time and consequently better chance there has been for its transportation by accidental agencies.

Like all authors who advocate union of the Galapagos Islands with the mainland, Dr. Scharff does not distinguish between a union of the islands with each other, which is geologically probable and is an almost unavoidable conclusion from a study of the fauna, and their union with the mainland, which is highly improbable on geological and physiographical grounds, and is not merely unnecessary to explain the fauna but impossible to reconcile with its peculiarities by any reasonable theories which take into account all of the consequences of such union.

While Doctor Scharff's interpretation of the data is based upon fundamentally different principles above noted, and his statements as to fossil distribution are often inaccurate or incomplete, yet the numerous distributional data which he presents of modern invertebrates are of great interest, and, if interpreted along the lines which I have used, they fall completely into line with the vertebrate evidence. We cannot usually indeed check the conclusions drawn from modern distributional relationships by the fossil record. Many groups are altogether unknown, and the record in others is very scanty, but the same general relations clearly apply. The survival in Western Europe on one side, in southeastern North America on the other side, of a somewhat primitive cycle of Holarctic distribution; the survival in the Mediterranean region on one side, in Central America and the Antilles on the other, of a more primitive cycle; of a still more primitive cycle in Africa and South America; and the progressively greater amount of divergent or parallel specialization in the survivors of the earlier cycles; the antique and fragmentary character of the faunæ of the oceanic islands, progressively more so in proportion to their smallness and isolation—all these conform to the vertebrate distribution. And with invertebrates as with vertebrates, every year adds to the number of the types which, while now limited to the peripheral continents and oceanic islands and highly discontinuous in their range, are shown to have inhabited formerly the central Holarctic region. It appears that many, one might perhaps say most, invertebrates are more readily transported across ocean barriers than vertebrates, especially mammals, even making due allowance for their greater antiquity. This also we should expect.

I do not think it necessary to catalogue the errors or inaccuracies in presenting the evidence afforded by fossil vertebrates. Such errors are unavoidable in a subject of so broad a scope, and excusable enough, if they do not lean too much to one side. I shall cite but one instance, and this in justice to my distinguished confrère Professor Depéret. Doctor Scharff concludes his summary of the North American records of the Evolution of the Horse with the following remarks:[133] "And yet not a single transition from one genus to the other seems to be known. No wonder that one of our foremost paleontologists exclaims, 'The supposed pedigree of the horse is a deceitful delusion, which simply gives us the general process by which the tridactyl foot of an ungulate can be transformed in various groups into a monodactyl foot in view of an adaptation for speed, but this in no way enlightens us on the paleontological origin of the horse.'" Such a statement, coming from so excellent an authority, seems startling until one verifies the quotation and finds that it refers, not to the American records, but to the ancestry of the horse as presented in Gaudry's[134] Enchaînements, to the European series *Palæotherium, Anchitherium, Hipparion gracile* and *Equus*. Depéret takes care to premise that he is speaking only of this European series, and while I think the criticism goes too far—it should at least be modified by changing "ungulate" to "perissodactyl" in view of what we know about the Litopterna—yet the criticism is largely justified in its proper context. As applied to the American series it is altogether unwarranted.

---

133. *Op. cit.*, 147.

134. A. Gaudry: Enchaînements du Monde Animal 3: Mammifères tertiaires. 1878.

REPRINTED FROM

ANNALS OF THE NEW YORK ACADEMY OF SCIENCES

VOLUME XXVII, PAGES 1-15

JANUARY 25, 1916

SOME REMARKS UPON MATTHEW'S "CLIMATE AND EVOLUTION"[1]

BY THOMAS BARBOUR

WITH SUPPLEMENTARY NOTE

BY WILLIAM DILLER MATTHEW

"Climate and Evolution," which is really more than its title would imply an essay on the origin and dispersal of vertebrate life, appeared in February, 1915, from the pen of Dr. W. D. Matthew.[2] It is by far the most scholarly and carefully constructed essay of its kind which has appeared and it demands a careful reading by all who take interest in perhaps the greatest of biological problems—the why and wherefore of the dispersal of animal life as we find it to-day and the past history of present conditions.

Matthew's thesis, in a few words, is that the permanence of the continents and ocean basins is a surely established fact, that cyclical climatic change has been the principal known cause of the present distribution of land vertebrates, and that this distribution has been effected by successive southward migrations from a Holarctic center of dispersion, and that the impetus for these migrations is to be found in the theories of the "Alternations of moist and uniform with arid and zonal climates, as elaborated by Chamberlin." There is small occasion for me to review or criticize the great bulk of evidence which Matthew has presented, specially where he has drawn upon his profound knowledge of recent and fossil mammals. In the main his contentions are highly convincing, especially where he

1. Presented by title before the Academy, December 13, 1915. Manuscript received by the Editor October 22, 1915.

2. Ann. N. Y. Acad. Sci., vol. 24, pp. 171-318. 1915.

also draws conclusions from the mammals, a group for which geologic record is adequate in comparison with the fragmentary evidence regarding the history of recent birds, recent reptiles and amphibians. With some of these groups, as, for instance, hylids and cystignathids, it is hard to rid oneself of the belief that their origin was Antarctic and not Holarctic, for the northern outpost species seem to be so obviously the depauperate offshoots of the elaborate southern stock. Matthew, however, would argue by analogy with mammalian evidence that these species are reëntrants into the area of origin, and that the great result of speciation which we see now in the southern headquarters of hylids and cystignathids shows that this region was peripheral in relation to their area of origin. The steps of reasoning whereby Matthew arrives at these conclusions are carefully presented in his essay and there is no object in recounting them here.

It is of the general question of land bridges and of the relation which some islands bear to continents that I have been thinking for some time, and it is only because I have had some field experience and have given thought to these matters that I have the temerity to take issue with Dr. Matthew, knowing full well that many will maintain that his opinion outweighs mine—a possible assertion I am by no means ready to deny.

I take exception to statements such as this, where in speaking of land bridges (page 10) Matthew says, "I can see no good reason why the only animals which availed themselves of such continental bridges should be the ones which might be accounted for in other ways, while those which would furnish conclusive proof *are invariably absent*." (Italics are mine.) I have maintained elsewhere that a waif fauna is easily recognized as such, and that the presence of burrowing amphibia, onychophores, cyprinodont fishes and many other groups of delicate organisms which are balanced to one particular environment cannot by any stretch of the imagination be distributed by "flotsam and jetsam" methods; and further that the element of the vast extent of geologic time does not in any way affect the probability of such dispersal, since it cannot be supposed *ever* to occur.

Again, on page 18, we read that Austro-Malaya is the debatable ground between the Oriental and the very distinct Australian region; but that the consensus of opinion classes it by preference with the Australian. It includes, we are told, Celebes, the Moluccas, Timor and the smaller islands and is separated from the Oriental region by "Wallace's Line." This is surely a step backward, for "Wallace's Line" marks the limit of but a small fraction of the whole species total of the Indonesian fauna, while the area from the Lesser Sunda Islands and Celebes on the one hand to Papuasia on the other represents a great transition zone, where a dominance of Malayan types may be found in the western part which merges into a predominance of Australian types in Papua. There is no real boundary line in the entire area and no reason to expect one.

Again Dr. Matthew in his "Summary of Evidence" (page 139) states that "the continental and oceanic areas are now maintained at their different levels chiefly through isostatic balance and it is difficult to believe that they could formerly have been reversed in any extensive degree." Then, on page 140: "A rise of 100 fathoms would unite all the continents and continental islands, except perhaps Australia, into a single mass, but would leave Antarctica, New Zealand, Madagascar, Cuba and many smaller islands separate." These four areas Matthew believes to have been always isolated islands, and if we can show a probability that any one of them was continental, we can at least make more reasonable a proposition that they all were once united to some other continental land. This point will be returned to later on.

Now a word regarding isostasy. There is hardly a principle in geology concerning which there is greater uncertainty among geologists than the matter of isostatic balance. Only one thing is sure, isostasy must meet and conform to known or presumably known facts, and the fact that fundamental changes have taken place in the form of the earth's surface in recent geologic time is not to be denied. Such features as the Great Rift Valley of Africa and its continuation, the Red Sea and the Dead Sea, the Black Sea, the Basin of the Mediterranean, are held now by geologists to be the results of nothing but gigantic and not at all ancient downthrown fault-blocks. For other examples of changes of land and sea level with relation to each other, the Valley of the Po and the Central Valley of California are good evidence. The argument of isostatic balance may probably be held to control the conditions in the Pacific Basin as a whole, but isostasy cannot be used effectively as an argument in a relatively small area anywhere. Professor R. A. Daly tells me that there is clear evidence of the fragmentation of a great land mass, including the Fiji Islands and New Caledonia, but that there is no evidence known at present of such a condition outside of a line joining Yap, in the Caroline Islands, the Fijis, Kermadecs and New Zealand. Besides this radiolarian ooze has long been known from Barbadoes, Trinidad, Aruba, Buen Ayre and Curaçoa, supposedly only to be derived from the deep sea, but the origin of this series of deposits has been somewhat in dispute. Two recent papers by Dr. G. A. F. Molengraff, however, describe deposits of which there can hardly be any question whatever; one is "On Oceanic Deep Sea Deposits in Central Borneo,"[3] while the other is entitled "Over mangaan Knollen in

3. Kon. Ak. Wet. Amsterdam, Reprint from Proc. of meeting June 26, 1909: 141-147. [Reprint: 1-7.]

mesozoischen diepzeeafzettingen van Borneo, Timor en Rotti, hun beteekenis en hun wijzer van Opstaan."[4] These papers show that on the islands of Borneo, Timor and Rotti, at an elevation of about 4000 feet, very extensive deposits occur which a microscopical examination shows to be composed of radiolaria, together with the manganese nodules so characteristic of the deep sea. In other words, Molengraff has found an extensive area of deep sea floor raised to 4000 feet above the present sea level. On the southeast coast of Africa, W. M. Davis noticed the truncation by the present shore line of extensive concentric terraces, traceable far inland, which could only mean the down-faulting of a gigantic block of material to bring the shore line into its present state. It will be said at once that some of these changes of level have taken place in zones known to be in incomplete isostatic adjustment, but this is a matter of no moment whatsoever in comparison with the fact that change of level may be found to have occurred in the very areas where the islands under discussion are to be found. Celebes does not lie upon the continental shelf and yet the island has an obviously continental fauna, and Dr. Matthew has told me himself that Celebes has been a source of no small worry to him. Cuba has similarly a large fauna, derived from the American continent, although it does not lie upon the continental shelf. Vaughan, a thoroughly conservative observer, believes (in litt.) that Cuba was quite possibly separated, by the down-faulting of blocks of material, from both Haiti and the mainland. Dr. Matthew (in litt.) says: "The fault block theory is of course a very familiar one; its application to continental movements is undoubtedly extensive, although it is just now somewhat of a fetish among stratigraphers, as folds were fifty years ago. But on land the great fault blocks are largely compensated by erosion, so that they do not involve so extensive a displacement of adjoining surfaces as one might at first suppose. Their application to explain submarine conditions where such compensation does not occur brings them into an apparent conflict with isostatic adjustments. Considering that we cannot possibly *prove* their responsibility for the sudden changes from shallow sea to abyssal depths in any case, I am inclined to avoid hasty ascription to such features as block-faulting. I have passed beyond the stage of immaturity when one is unreasonably certain about things." I can only add that I am as far from being unreasonably certain regarding isostatic adjustments in general as Dr. Matthew is regarding marine down-faults. To the zoölogist these geologic problems seem so differently interpreted by different and equally gifted and trustworthy students that one is inclined to relegate them all to the limbo of where "you pay your money and take your choice."

4. Kon. Ak. Wet. Amsterdam 23: 1058-1073. [Reprint: 1-16.]

Vastly different, however, is the matter of the zoölogic evidence presented by the faunas of some islands as indicative of the island's geologic or geographic history. Dr. Matthew lays great stress upon the importance of the mammalian element in the fauna. Here a word of caution is not amiss, for mammals act queerly upon islands and often have a way of being most strangely absent, as this is the group which has greatest difficulty in surviving in a limited area. Trinidad, a large island separated by a very narrow strait from Venezuela, has a reasonably full quota of mammalian inhabitants, while the large and heavily wooded island of San Miguel, just off the coast of Panama, has but a few small mammals, quite a contrast to Coiba Island, farther north off Honduras, where even a peculiar deer is known still to occur. Gorgona Island, off Colombia, also with luxuriant vegetation, has a peculiar *Cebus* (*Cebus curtus* Bangs), a peculiar *Proechimys,* and so far as known no other mammals. Yet these differences are all among islands on the shelf and near or fairly near the shore; and I could multiply the examples!

Now I do not believe, with Matthew, that the Antilles are oceanic islands—islands which have received their fauna by fortuitous transportal. My reasons for thinking as I do are these: first, I believe that the islands of the Antillean chain have too evenly distributed and homogeneous a fauna for it all to have been fortuitously derived; secondly, I consider the fauna to be composed of too many different animal phyla; and thirdly I believe that many of these elements are not of a nature to have withstood "flotsam or jetsam" dispersal. We must now consider Matthew's exposition of the natural raft hypothesis (page 37 *et seq.*). He states: "1) Natural rafts have been several times reported as seen over a hundred miles off the mouths of the great tropical rivers such as the Ganges, Amazon, Congo and Orinoco. For one such raft observed, a hundred have probably drifted out that far unseen or unrecorded before breaking up." This is obvious and undoubted. But, and this is most important, these rafts, even the very large ones, float low in the water; they soon become soaked with salt water in a calm sea, rippled over or broken over if the sea be choppy or rough as it is in the trade wind or monsoon belts. Only organisms or their eggs which are encapsulated or otherwise naturally resistant can withstand these conditions. Molluscs are stimulated to activity by dampening, but most are killed by salt water—although some such as *Cerion* are resistant. Scincs and gekkos show by their distribution that they may be carried about in this way. Amphibians, amphisbænians, naked gastropods, earthworms, fresh-water fishes or crayfishes, *Peripatus* and a host of such delicate creatures simply cannot withstand salt water. No such creatures have ever been observed upon any raft, of the very few recorded, and to transport cyprinodonts, ampullarias and the host of other fresh-water types one meets with in Cuban ponds, for instance, the raft would have to include a puddle, at least, of fresh water. Supposing that an amphisbænian, to take a good example, withstood an ocean voyage upon a raft, how would the landing take place? The raft would have to make a haven and then ground in such a way that a very delicate, blind and legless lizard would be enabled to reach a suitable environment on shore. I only ask the reader to tramp West Indian shores with this in mind. If a pair or a gravid female did not make the voyage the process would have to be repeated promptly. Now consider the number of rafts each of which would have to carry an amphibian or a pair of them and which would have to start on their journeys before one would reach shore so as to permit a landing such as I have indicated. Think of the number broken up at sea, and the still greater number broken up on a tropic beach—where the sun would instantly kill crawling amphisbænians—and we see at once how excessively improbable is a single occurrence such as this. But the important point is that five West Indian Islands support peculiar amphisbænians; two species occur upon Cuba, two others related to these two on Haiti and two others similarly related to the Haitian types on Porto Rico, while but one type is as yet known upon St. Thomas and Sta. Cruz. To account for the presence of these creatures, then, eight practically inconceivable voyages must be postulated, and I have only cited one improbability out of many hundred necessary to derive all the organisms practically or wholly incapable of such sea travel and which are found in the Greater Antilles. A few such cases as Amphisbæna settle the status of the greater Antillean fauna to my mind absolutely, paucity of mammals and possibly disputable geologic evidence to the contrary notwithstanding.

Matthew's second premise is: "2) The time of such observation of rafts covers about three centuries (I set aside the period of rare and occasional exploring voyages). The duration of Cenozoic time may be assumed at three million years (Walcott's estimate)."[5] But is it not true that this multiplication of time or any other, of course, affects only the number of rafts and does not in any way alter the resistance to raft conditions of the creatures which I have already chosen above as examples or the possibility of their being able to swim. It really carries no weight in this connection.

Matthew's third point: "3) Living mammals have been occasionally observed in such records of natural rafts. Assume the chance of their occurrence (much greater than of their presence being noted) at one in a hundred." We readily agree to the assumption and know that during the few years of human observation rafting mammals have been observed.

[5. This estimate has been revised, upon the basis of Barrell's studies, to sixty million years. See page 37 of the present volume.—E.H.C.]

It occurs to us, however, that multiplying the three hundred years' time of human voyages by the ten thousand necessary to occupy even the short Cenozoic period and then with this condition met, we find mammals *infinitely rarer* upon all islands than they should be if rafted according to Matthew's postulate. Similarly we find many of the reptiles most capable of withstanding transport by rafts conspicuously absent in the West Indies; for example, *Basiliscus,* almost semiaquatic, is absent in the Antilles, as *Varanus* is absent upon Madagascar: they were possibly derived in common with many other of the islands' original continental inhabitants and have failed to survive. We know that some edentates and a few rodents have become extinct in the West Indies, some within the last half century; why, in some cases, we cannot guess; probably more have gone the same way, of which we have unfortunately no fossil remains. So also the boa-constrictor, which has been observed to be carried once to St. Vincent from the Orinoco and which has probably come comparatively often, has never succeeded in establishing itself. Why some types fail to survive upon islands while others of apparently similar habit flourish is one of the enigmas for which no mite of answer has appeared. Matthew's arguments, which he numbers 4 and 5, may be considered together: "4) Three hundred miles drift would readily reach any of the larger oceanic islands except New Zealand. Assume as one in ten the probability that the raft drifted in such a direction as to reach dry land within three hundred miles. 5) In case such animals reached the island shores and the environment afforded them a favorable opening, the propagation of the race would require either two individuals of different sexes or a gravid female. Assume the probability of any of the passengers surviving the dangers of landing as one in three (by being drawn in at the mouth of some tidal river or protected inlet), of landing at a point where the environment was sufficiently favorable as one in ten, the chances of two individuals of different sexes being together might be assumed to be one in ten, the alternate of a gravid female as one in five. The chance of the two happening would be $1/10 + 1/5 = 3/10$. The chance of the species obtaining a foothold would then be $3/10 \times 1/3 \times 1/10 = 1/100$." He then continues, "If then we allow that ten such cases of natural rafts far out at sea have been reported, we may concede that 1000 have probably occurred in three centuries and 30,000,000 ((200,000,000)) during the Cenozoic. Of these rafts, only 3,000,000 ((2,000,000)) will have had living mammals upon them; of these only 30,000 ((200,000)) will have reached land, and in only 300 ((200)) of these cases will the species have established a foothold. This is quite sufficient to cover the dozen or two cases of Mammalia on the larger oceanic islands.

"I have considered the case only in relation to small mammals. With reptiles and invertebrates, the probabilities in the case vary widely in different groups, but in almost every instance they would be considerably greater than with mammals. The chance for transportation and survival would be larger and the geologic time limit in many instances much longer. Wind, birds, small floating drift and other methods of accidental transportation may have played a more important part with invertebrates, although they cannot be invoked to account for the distribution of vertebrates. The much larger variety and wider distribution on inframammalian life in oceanic islands is quite to be expected. And the extent and limits of such distribution are in obviously direct accord with the opportunities for over-sea transportation in different groups."

In the estimate which Matthew has made there seems to be an obvious error, for should we postulate 1000 rafts in 300 years we would have 10,000,000 rafts during the 3,000,000 years of Cenozoic time, not 30,000,000 as Matthew has it, and the chances of the whole concatenation of events are reduced by 66 per cent.[6] But even this reduced estimate would if true bring more mammals to most islands than we find. Let us, however, for the sake of argument, admit that some mammals might be transported in this way, is the premise true that other creatures will be more easily carried? Some will, and these types by their haphazard occurrence can now be recognized easily; others most certainly will not. Matthew has not realized the enormous sum total of different species which go to make up the fauna of such islands as Cuba and Haiti. Such a vast number of species would require squadrons of rafts at frequent intervals, even if only ancestral stocks were transported from which many species arose after coming to the island by some sort of adaptive radiation. Another important point has also been missed. Almost all of these isolated groups of individuals have grown to be well differentiated island species. Distinct from the related forms of the mainland and neighboring islands, they represent types evolved in complete isolation; an occasional raft bearing individuals from the parent stock would by preventing breeding in, at least in some cases, prevent speciation by isolation taking place.

Let us for a moment consider the Antillean chain as a whole; it is utterly impossible that ocean currents could now or in the past have brought rafts with equal frequency to all parts of this island arc, and yet the same types reappear upon island after island all the way from Cuba to Grenada. Rafting from island to island could certainly not have occurred, since there could never have been large rivers on them had they always re-

[6. This error was recognized and subsequently corrected by Dr. Matthew. See pages 37, 159 of the present volume.—E.H.C.]

tained their present size. The fauna is far larger in number of species upon the Greater Antilles than upon the Lesser, as the conditions favorable for the survival of species are obviously better upon the large islands with their luxuriant vegetation than upon such barren islets as, for example, Sombrero or Redonda. The types, however, which have been able to survive upon Sombrero or Saba are just those which are found, along with many others, upon Cuba or Haiti. In my "Herpetology of Jamaica" and "Zoögeography of West Indian Reptiles," I have gone into this matter in detail and there is no need of repeating what has been said there. This homogeneity of the fauna is the best possible proof that winds (tornados, hurricanes, etc.), birds, small floating drift, etc., have played no considerable part in populating the island by carrying eggs or adults, since it is inconceivable that by these means the same improbable choice of passengers would be carried to so many islands.

Matthew, upon the basis of the mammalian fauna of Madagascar entirely, he believes, derivable from a few waifs, and from the fact that the island is not upon the present continental shelf, concludes that it is an oceanic island. We may grant that all the lemurs have radiated from a single type, and this may have been a waif type—all this for the sake of argument—but what does the rest of the fauna show? We find abundant amphibians of many different families, as well as a great host of other land and fresh water organisms which cannot by any stretch of the imagination be considered as more probably capable of surviving raft transport than mammals, nor in very many cases of possibly surviving such transportation at all. Yet such types as these are most abundant upon Madagascar, in individuals and in species—species representing wholly unrelated mainland stocks and not those which might possibly have arisen after coming to the island.

My friend Dr. G. M. Allen has contributed the following note regarding Madagascar which is interesting in this connection. He writes me: "The total absence from Madagascar of any native species of the typical Murinæ seems to be a striking bit of negative evidence against a chance population of the island. All the nine genera of indigenous rodents are cricetine in their affinity, though now considered to represent a special subfamily by themselves—Nesomyinæ. The cricetine-like rodents are abundant still in the Americas, less so in number of species in northern Eurasia. The African Lophiomys is nearly related. If we consider the more specialized typical Murinæ as representing a later development of the Muridæ, it is easy to account for their absence from America, if formerly, as now, their northward range did not extend to the East Siberian region, whence they could have crossed by land bridges if such existed.

That no member of so widespread and successful a type in the Old World as *Mus* (in the broad sense) has reached Madagascar, it seems evident that it is because none have been able to cross the intervening water. If nine distinct genera of cricetine-like rats or their ancestor or ancestors could have reached Madagascar by chance methods, it seems inconceivable that no single Murine could have done so, despite the great adaptability and abundance of the representatives of this group. The most attractive explanation of this fact is that the more primitive cricetine-like rodents reached Madagascar by land connection from Africa and that they were subsequently isolated there before the advent of the more specialized and successful Murinæ, which have now totally replaced them on the mainland."

So much for the question of rafting—some creatures can be carried and some—many more—cannot survive such conditions. We may recognize by their haphazard distribution and by their habits in the field those waifs which can withstand raft transport. Yet even these resistant types are very often strangely absent; there are no varanids in Madagascar, and yet we should naturally suppose that they would be among the very first immigrants by raft carriage.

This whole question is really but a side issue with Matthew. He is far more absorbed in other problems; hence it is only fair to say that this island question is of secondary interest to him. Dr. Matthew's masterly handling of his chapters dealing with mammals is beyond praise. He has surely shown that the present distribution of most if not all of the recent mammal groups may be plausibly explained without having recourse to postulating extensive changes of land forms. But Matthew deals with some other matters as one without authority, and one feels that his opinions would be different had he seen and not merely read about the rafts, and the landings of the rafts, of which he must perforce write to explain his ideas.

Many will notice trifling inaccuracies in the text, such as the statement that the large ground-birds of modern times are "to-day peculiarly inhabitants of arid regions." There is the New Zealand kiwi in the rank damp fern forest and the host of different cassowaries in Ceram, the Papuan Islands, New Britain, Queensland, and anyone who has ever tried to hunt cassowary knows how well they are adapted to getting about in the densest jungle in the world. However, such points are of so small import that it is hardly worth while mentioning them. My final word is not to advise but to adjure everyone who aims at a wider knowledge of natural history to read Dr. Matthew's paper.

SUPPLEMENTARY NOTE[7]

By W. D. Matthew

Before replying to Dr. Barbour's criticism I will say that it is of the sort that is peculiarly welcome, not merely because of its courteous and considerate tone, but because of the author's wide field experience and knowledge of the practical conditions and circumstances of environment that govern the probabilities of any theories of distribution. However we may differ in our interpretation of the evidence, we agree in emphasizing these factors, in the impossibility of solving such problems by the study of any one group of animals, and especially in the need for securing more complete distributional data, towards obtaining which he has devoted so much time and energy.

Many a false theory gets crystallized by time and absorbed into the body of scientific doctrine through lack of adequate criticism when it is formulated. I should be very sorry to see that happen to the views that I have maintained and hope that adequate and competent criticism will serve to sift out truth from error before they are either adopted or discarded.

Concerning the question of isostatic adjustment I do not think it necessary to make any especial comments. I stated in the outset that I was applying the geological views set forth by Chamberlin and others. The evidence that Barbour cites was known to me and is covered by the qualifying phrases that I used; but the general discussion of the permanence of ocean basins I shall continue to avoid.

Dr. Barbour devotes considerable space to a criticism of my attempt at a statistical presentation of the possibilities of "rafting." He is indeed far less critical of its assumptions than I am. I tried to make it clear that its sole purpose was to show that the hypothesis involved not "miracles of transportation" or infinitesimal chances but reasonably probable chances. The quite inexcusable error in my figures to which he calls attention is not really material, nor is his conclusion that the calculation shows too many mammals for the known or inferential instances (I do not see that it does, by the way; but as the point is immaterial will pass it by). The main reason for introducing a calculation based upon a series of highly inexact approximations was to determine whether this method of transportation would afford a reasonably probable alternative to continental connection, in accounting for the presence of mammals on certain oceanic

7. Manuscript received by the Editor November 12, 1915.

islands, in which it appeared impossible to explain the characters of the mammal fauna upon any reasonable hypothesis involving continental union.

When it comes to applying the rafting theory to lower vertebrates and invertebrates, Dr. Barbour is quite right in insisting that one should have a special knowledge of the particular habits of each group in order to judge of possible or probable methods of its transportation. Because I lack that special knowledge and field experience I avoided discussing it. I do not know how far the rafting theory is applicable to account for their distribution, and how far it may be more reasonably explained in other ways. But I do notice that Barbour appears to consider only the transportation of the adult animals, and says almost nothing about the possible transportation of their eggs. It is just because mammals do not lay eggs that their presence in oceanic islands seems peculiarly difficult to account for, save through former union with continents. Wallace, it will be remembered, for this reason regarded their presence as a dependable criterion of such union. Like most students of geographic distribution, I started from Wallace's views as a basis, and if I have modified them in an opposite direction from many of my confrères it is because I found that in certain particulars they did not fit the details of distribution, past and present, in the groups with which I am best acquainted. If mammals laid eggs, and especially if the eggs were numerous and of very small size, I should find it far easier—indeed quite too easy—to account for their presence on oceanic islands; the difficulty would be to account for their general absence. It is because they do not that I was compelled to discuss a rather complex hypothesis to account for their presence in Cuba, Madagascar and elsewhere. There are numerous other accidental means of transport which might be and have been invoked to account for dispersal of lower animals, but the ones I discussed were the only ones that, so far as I could see, would be possible for mammals. And for very large terrestrial mammals these seemed to me physically impossible. But no such mammals closely allied to continental ones are found on any oceanic islands;[8] those remotely related are derivable from the much smaller and more generalized ancestors which we find living in older epochs on the continents, and it is these small common ancestors whose dispersal must be accounted for.

It is the above considerations that underlay the remark to which Barbour takes exception on page 150. I do not think his objection is war-

8. There are certain exceptions to this statement—Malia, Cyprus, Crete and Celebes. These exceptions do appear to call for continental union.

ranted, for while I do not question his judgment as to the improbability of transporting certain types of adult reptiles, amphibians or invertebrates in the particular way that I have used to account for certain peculiarities in mammalian distribution, yet the transport of their eggs is a much less serious difficulty, and there are several other possible methods. These have been extensively discussed by others and I need not go into them. While it does not seem physically possible, for instance, that a mammal could survive being caught up by a hurricane or tornado, carried a long distance and dropped, I do not see any particular *impossibility* in the transport of young or eggs of amphibians or reptiles or invertebrates by such means. Indeed there is considerable evidence along this line on record. And if it be not a physical impossibility, then the element of geologic time *does* enter into the question of its probability.

So far as the West Indies are concerned, I do not at all suppose that no one of them has ever been united to any other. All are in lines of great disturbance and uplift; several are united by shallow platforms, and I see no reason at all against accepting Vaughan's view that Cuba and Haiti were probably united during the Tertiary, although now separated by a deep water channel. The mammalian evidence would accord well with that view, and recent discoveries in Porto Rico rather suggest that that island, as well as the Bahamas, may have been included in the union. But the mammalian evidence seems to me distinctly against any direct union with either North or South America in the middle or later Tertiary, and fairly conclusive against any such Pleistocene union. Indirect union with an isolated Central American or Floridian island serves to raise problems of dispersal more difficult than any that it solves, because we have no past distributional evidence to go upon. It is too speculative to be worth discussion.

Dr. Barbour's criticisms deal, as he very cordially and generously insists, with a side issue of the main discussion in my paper. I may add that it is a side issue concerning which I have no desire to be dogmatic or positive.

My statement that the large ground-birds are to-day peculiarly inhabitants of arid regions should have been more carefully qualified. It referred especially to the three large forms—the ostrich, rhea and emu—characteristic of the arid interior of the three southern continents. I had not meant to say that there was any such uniform association of habitat as the remark may seem to imply.

In reply to Doctor Allen's argument concerning Madagascar cricetines, the following points may be considered:

(1) All the Malagasy rodents belong to a peculiar group of cricetines,

the Nesomyinæ; that is to say, if the classification is a natural one they are descended from a single stock, a single ancestral type, and as the group is peculiar to the island the inference would be that its differentiation took place after it reached the island and not before. The arrival of that type represents therefore one invasion and not nine.

(2) The cricetines are the older group, prevalent in Holarctica since the beginning of the Oligocene and presumably widely spread in the Ethiopian and Oriental regions during the Middle and later Tertiary. The murines are a much later development and comparatively recent immigrants into the tropical Old World, where they have supplanted the older cricetines more completely than in the less accessible tropical New World. The cricetines have therefore had a much longer time than the murines to reach Madagascar through accidental transportation; it is to be expected that the single stock which arrived should be a cricetine.

(3) The Malagasy lemuroids, Gregory has recently shown, all belong to a single group, distinct from any of the continental lemuroids, and may equally be supposed to represent a single invasion. But the wide differentiation indicates that this invasion must be a much earlier one than that of the cricetines.

(4) The Carnivora—all viverrines—are not so clearly derivable from a single stock, but they may be so (*Cryptoprocta,* although highly specialized, is fundamentally a Middle Tertiary viverrine). The date of arrival can hardly be much later than Oligocene, and certainly not earlier.

(5) The insectivores—all Centetidæ—are likewise derived from a single stock, peculiar to the islands, its only continental relative being the potamogalids. The diversity of the Malagasy centetids indicates an early Tertiary invasion.

(6) The pigmy hippopotamus must be a late arrival, for the hippopotami were not evolved until the late Tertiary, and no adaptive radiation has occurred in Madagascar.

(7) An Eocene land connection and subsequent isolation might account for the lemurs and centetids, but would not allow of land invasion by viverrids, cricetines or hippopotami. We should expect also that at least some remains of the Eocene ungulates and creodonts would survive, since they were not displaced by higher types as on the mainland.

(8) An Oligocene land connection with isolation before and after might explain lemurs, centetids and viverrids, but not the cricetine group and certainly not the hippopotami, and would be difficult to reconcile with the absence of monkeys and of the primitive ruminants and proboscideans (all already in Africa in Lower Oligocene).

(9) A Miocene or Pliocene land connection with isolation before and after would explain the cricetines, and either the viverrines (if early) or the hippopotami (if late), but hardly both. It would be irreconcilable with the presence and diversity of the lemurs and centetids, the absence of monkeys and of the rest of the later Tertiary mammal fauna of Africa.

(10) A land connection throughout the Tertiary would account for the presence of the few mammal types that have reached the island, but not for the complete absence of the very numerous groups of several successive faunas which are not present. The amount and character of the adaptive radiation among the half dozen stocks that are present indicates the absence of monkeys, ungulates, cats, etc., in the past as well as in the present, as also that the half dozen different elements of the mammalian fauna were due to invasion by single types at several different epochs. The indicated epoch corresponds in each case to the time at which the geologic record indicates that their ancestral stocks were prevalent on the adjacent mainland.

(11) Madagascar, it should be remembered, is an island of almost continental size and great geologic antiquity. It is not a transitory islet, such as Barbour cites in support of his contention that mammals behave strangely on oceanic islands. If a continental fauna once gained access to it there would be no reason for its extinction, nor opportunity for the expansive radiation of a few stocks. If Doctor Allen's explanation were correct, the relations of the fauna to the Ethiopian should be similar to those of Borneo or Sumatra to the Oriental mainland fauna, save for a wider amount of differentiation from the late Tertiary and Pleistocene invaders of the Ethiopian region. The briefest summary of its character, such as is given above, is enough to show how widely it differs from that type.

NOTE ON THE WEGENER HYPOTHESIS, ETC. SUPPLEMENTARY TO "CLIMATE AND EVOLUTION"[1]

By William Diller Matthew

Wallace in one of his earlier works[2] observes that the old idea of the reasons for the present distribution of animals and plants attributed it entirely to suitable climate and environment. He shows, however, that this explanation is quite inadequate. It fails to account for the great differences in the fauna between regions of similar climate and environment, such as the East and West Indies, tropical America, and tropical Africa, or between the two great faunal divisions of the East Indian islands, one allied to the Oriental mainland fauna, the other to the Australian mainland fauna although a broadly similar climate and environment prevails throughout the whole archipelago. He proceeds to show how the past and present geographic relations, the land connections and water barriers that exist to-day and may be inferred to have existed in the past, provide an explanation of the faunas peculiar to each region or locality not explainable by suitable environment.

Wallace's work is representative of a great literature on distribution, past and present, which has accumulated partly before, chiefly since his time. In his subsequent contributions, especially the great monograph on Geographical Distribution, and the later volume on Island Life, he became distinctly more conservative, and was more and more inclined to caution in inferring land connections and ocean barriers, and to give more weight to the accidents of oversea transportation.

The science of zoögeography has been greatly developed since Wallace's time. A great mass of exact knowledge of the present distribution of animals and plants has been accumulated, and our knowledge of their past distribution while still relatively scanty is vastly greater than in 1860-1880. Geology too has advanced enormously and the data on which our theories and interpretations rest are so vast that no one man can really understand

---

[1. The following notes were left in manuscript form at the time of Dr. Matthew's death, and were most kindly presented for publication in the present volume, by Professor Charles L. Camp of the University of California. Professor Camp makes the following statement with regard to this new contribution: "I have no idea as to when this was written except that it is on a writing pad which was bought in the Orient, so that it must have been during his visit to India. This is all the more likely since he starts the article with some remarks about the East Indies."—E.H.C.]

2. Alfred Russel Wallace: Proc. Linnæan Soc. London 4: 172-184. 1860.

the significance of more than a small part of the evidence. Yet in order to interpret it one needs not merely a knowledge of the data, not merely a series of lists of fauna but a thorough comprehension of what these data mean, what the faunal differences may or may not signify, what the geology and physiography and bathymetric data indicate as probable or reasonable and what they render in varying degree improbable. Overwhelming faunal evidence might override weak geological evidence, and *vice versa*. But a faunal list, and especially a list of fossil fauna may mean much or little, according to the accuracy and certainty of the identification of groups or species listed, the probable completeness of the record and the faunal facies that it represents.

The non-occurrence of marsupials in the record of later Tertiary faunas of Western Europe is a significant fact in the distribution of this order. Their non-occurrence in the earlier Tertiaries of Asia has little weight, and until recent discoveries had none whatever. For the later Tertiary faunas of Western Europe are extensive and varied, representing various facies and as a whole probably most, though by no means all, of the mammals then living in that region. One might fairly expect that if any marsupials survived there until the later Tertiary, some remains of them would have turned up among the known fossil faunas. On the other hand the early Tertiary (Eocene) mammal record in Asia was until recently a complete blank, and is now known only from two areas, Mongolia and Burma. The Mongolian Eocene appears to be a fauna of plains or open country and presumably temperate climate (45° N.) and the Burmese Eocene apparently is a forest fauna, presumably tropical (18° N.). Neither is sufficiently well known as yet to infer that marsupials were not present in the region because they have not been found in the fauna collected; one can say only that they were certainly not dominant and probably rare if present on the Mongolian plains, and that they were probably not common in the Burmese forests. To infer their entire absence from the whole Asiatic continent from their non-occurrence in the Tertiary record would be going far beyond what the nature of the evidence warrants. It might indeed be inferred upon other grounds, from the broader relations of the marsupial and placental mammals, that the metatherian stage of evolution dominant in the Mesozoic mammals was gradually replaced in the northern world by the higher placental stage towards the end of the Mesozoic, and the marsupials, a surviving group representing this stage, still present in Asia in the late Cretaceous and Paleocene (of Mongolia) retreated southward about the beginning of the Tertiary and were absent subsequently save perhaps in the Oriental region and rare or absent there except in the now insular portions where they still survive. Or that they became completely extinct even in the East Indies and that the modern phalangers, etc., are invaders from Australia. Either inference would be in accord with the fossil record, but certainly has no claim to be proved thereby. For it is quite certain that the two local faunas of which we have only a small part preserved as fossils, are far from being the whole of the varied faunas that inhabited Asia during the early Tertiary. If our theories demanded it, it would be quite reasonable to suppose that marsupials lived in the Tertiary forests of Asia although they are not in the fossil record. The positive evidence of their occurrence in the Cretaceous and Paleocene of Mongolia suggests at least that they survived to a later date in the Oriental region. It does not prove it, nor does the Burmese record disprove it.

The fossil record presents us also with very varying values in its evidence according to the completeness of specimens and the certainty of their identification. There is a long record of erroneous identifications based upon fragmentary material or upon hasty or uncritical study. Many of the errors in the older literature have been corrected. Many more remain uncorrected, or the corrections have been ignored by subsequent writers. We no longer record *Palaeotherium* in Nebraska, nor *Anoplotherium* in Argentina. Yet many modern writers on paleogeography continue to record *Amphicyon* and *Listriodon* in the Argentine Tertiary faunas, and *Agriochoerus* and *Hyaenodon* in the Miocene of India. All these were based upon single teeth mistakenly referred to genera to which they were not even distantly related, the resemblance being a mere similarity in adaptation of that particular tooth. There are innumerable other instances of identifications erroneous or doubtful in varying degree among the fossil mammal faunas that I have had occasion to examine critically, and doubtless the proportion is equally large in other classes of animals and plants. For the most part they are not so wide of the mark nor so demonstrably wrong as these instances, but they are sufficient to lead astray the paleogeographer who depends upon published lists for his evidence and has no critical knowledge of its real nature.

It is in fact generally, I might almost say always the case, that these erroneous or doubtful records are the outstanding features of the evidence adduced in support of the more daring and sensational theories of paleogeography. The earlier writers adduced in chief support of their claims, evidence long since disproved and forgotten. Tapirs in the Orient and South America demanded a transpacific continent; camels in Africa and

South America, *Anoplotherium* in France and the Argentine, the southern distribution of ratite birds and zalambdodont insectivores and numerous other evidential data whose significance is now wholly different in the light of increased knowledge, were high points of the evidence for transoceanic bridges between various southern continents.

It is only fair to them to recognize that their interpretations were the obvious ones and to a superficial knowledge of the data or an uncritical mind, seemed the only explanation. Geologists, too, hardly emerged from the catastrophic period of their science, were far more ready to accept vast changes upon slight evidence than they generally are to-day. It is true that the greater lights in geology, as in zoölogy, the men who have survived to modern times, Lyell and Dana, Darwin and Wallace, were disposed to be conservative, and wisely hesitated to accept the plausible arguments of their more radical confrères.

Under present conditions it is impossible for anyone to cover so wide a field as did Wallace, and one must either depend upon compilations and lists in discussing the interpretation of faunæ, or restrict oneself to a relatively small group. It is of course true that the evidence of a single group, if correctly interpreted, will lead to the same conclusions as that from any and all other groups which are correctly interpreted. If a discrepancy appears, one or the other interpretation is wrong. But it will not do to depend here on the balance of evidence. The same fallacy may run through the interpretation of a dozen groups and be provably a fallacy in the one because of some method of checking up that can apply to it alone. This I think is very often the case in comparing conclusions drawn from the numerous groups of animals and plants of which the geological record is unknown or insufficient with the very few of which the geological record is adequate. Or again in comparing those groups whose simple anatomy affords numerous chances for a confusing parallelism or convergence to occur, with the groups whose more complex anatomy provides the means for more easy and certain recognition of that parallelism. Among fossil molluscs and brachiopods parallelism and convergence that extends to the whole external structure is common, and in certain instances convergence goes so far that phyla have been traced originating in the same genus, differentiating into distinct genera and then converging into the same genus again. Probably the same would be true of mammals if we dealt wholly or even chiefly with the external structure, but their complex internal skeleton provides a more ready means for detecting it, and it is hardly conceivable that a mammalian genus could be of such composite origin.[3] If, however, we had only the external parts of a marsupial flying phalanger and a placental flying squirrel, they might well be thought to be closely related, and no one without the testimony of the internal parts could prove that they were not. Similar instances could be multiplied indefinitely, and it is safe to say that one of the two principal reasons why mammals afford peculiarly valuable evidence in paleogeographic problems is that owing to their internal bony skeleton and particularly to their complex teeth, preformed and less readily modified by the environment than are the bones, and far less readily than the external form, we have preserved in them the best data for detecting parallelism and determining their true taxonomic relations. This is peculiarly true of the fossil forms as compared with most fossil invertebrates, and may well outbalance the comparative rarity and imperfect knowledge that we have of fossil mammals as contrasted with some groups of invertebrates.

The second reason is that the geological record of mammals is an extensive one, not indeed comparable in abundance and completeness with the record of a few groups of invertebrates, but better than the record of the lower vertebrates or of most groups of invertebrates. In fact little or nothing is known of the geological history of most invertebrate groups, including some which are special favorites in paleogeographic theorizing, and the temptation is strong to accept the most obvious and sensational interpretation of their present distribution, without pausing to inquire whether it conforms with the interpretation of groups that have similar present distribution but whose fossil record affords a means of checking its soundness. In dealing with such groups whose geological record is unknown or insufficient, an identical fallacy or group of fallacies is liable to run through the interpretation of all of them, leading to conformant but erroneous results.

In dealing with paleogeographic problems, the geologic evidence must be regarded as fundamental so far as it goes. Geology has progressed far since the earlier days of the science, and progress has made largely for conservatism. We have with us always, in this as in other sciences, a contingent of radical and speculative writers whose contributions loom large in popular science and are favorite subjects for expounding to undergraduates. But the working geologist, chiefly concerned in observing and interpreting the details of evidence in one area after another, is apt to take a tolerant but somewhat skeptical view of these more radical theories, as "playing for the gallery" but not contributing much to the progress of

3. It has been claimed, however, in the case of *Equus* by Abel.

a science that depends upon interpreting its data as the cumulative result of processes that are demonstrably in operation to-day, as they have been in the past history of the earth. It will not be questioned that the application of new concepts which can be tested and proven sound to the interpretation of geologic data, is a great step forward. But geology as popularly taught is full of more or less doubtful hypotheses, many of them of hoary antiquity, which never have been proved, and involve features difficult or impossible to reconcile with modern knowledge. Most of them probably if subjected to a really critical examination would share the fate of the old Nebular Hypothesis which after being taught for a century as the foundation of geologic theory has been set aside as mathematically impossible. Pending such examination one may view with an open mind the very ingenious and plausible theory of continental drift set forth by Professor Wegener and others (while not at all subscribing to his attempt to interpret the relations of land faunas, which is quite impossibly wrong). I cannot see how to fit it into the Tertiary mountain-making movements without involving radical changes in the relations of the Tertiary continents which run wholly counter to the evidence of the Tertiary mammal record. But I do not regard the Mesozoic or Paleozoic record of land life as competent to really prove anything much one way or another as regards the distribution of continents and oceans, and it may be that if a *vera causa*[4] were shown, Wegener's theory in some modified form (perhaps as initiating rather than causing orogenic uplifts) would be very helpful to Paleozoic and Mesozoic geology. But if there were no other difficulties, the mammalian faunas of Africa and South America, of North America and Europe, in their relations to Asia afford sufficient proof that these continents were in substantially the same relations at the beginning of the Tertiary as they are to-day.

---

4. By a *vera causa* I mean a mathematical demonstration that such a tendency to drift really does occur and can accumulate sufficiently in time to detach the continental masses from the anchorage imposed by friction. A wind may blow on a rock for a million years without the accumulated pressure moving it. But I can quite see that the analogy would not necessarily apply, although not mathematician enough to test it.

Reprinted From

Bulletin of the Geological Society of America

Volume XXIX, pages 657-666

December 30, 1918

AFFINITIES AND ORIGIN OF THE ANTILLEAN MAMMALS[1]

By William Diller Matthew

CONTENTS

1. Read before the Paleontological Society January 1, 1918. Manuscript received by the Secretary of the Society August 22, 1918. [A very similar treatment of this same subject will be found in the conclusions of Matthew's paper describing the results of work by the American Museum of Natural History in the West Indies. See W. D. Matthew: Proc. Am. Phil. Soc. **58**: 161-181. 1919.—E.H.C.]

LIMITATIONS AND RELATIONSHIPS OF WEST INDIAN MAMMAL FAUNAS, LIVING AND EXTINCT

In General

The indigenous land mammals of the West Indies consist of three groups: (1) Insectivora, (2) hystricomorph rodents, (3) gravigrade edentates. No perissodactyls (horses, rhinoceroses, tapirs, et cetera), no artiodactyls (peccaries, deer, antelopes, et cetera), no proboscideans (elephants, mastodons, et cetera), no true carnivores (dogs, cats, raccoons, mustelines, bears, et cetera). Nor are there any sciuromorph, lagomorph, or myomorph rodents, shrews, moles, hedgehogs, or opossums. All these large groups, most of them abundant and varied in Tertiary North America, are wholly absent. Nor do the Insectivora, rodents, or edentates include anything at all nearly allied to any North American members of the order or derivable from anything known to have inhabited North America in the later Tertiary.

Most of these North American groups invaded South America in the Pliocene and are part of its later fauna. In the Miocene and early Tertiary they are not found in South America, but their place is taken by a number of other groups. In place of perissodactyls, artiodactyls, and proboscideans were a number of groups of hoofed animals peculiar to Tertiary South America—the toxodonts, typotheres, litopterns, homalodontotheres, astrapotheres, pyrotheres. In place of the true Carnivora is a variety of marsupial carnivores (Borhyaenidæ) paralleling the true carnivores in structure and taking their place in the fauna.[2] All of these abundant and varied groups of ungulates and pseudo-Carnivora are lacking from the Antillean fauna, nor do the rodents represent more than two

2. A similar but distinct group of marsupial carnivores (Dasyuridæ and Thylacinidæ) developed in Australia in absence of true Carnivora and still survives there.

NOTE.—In this paper no account is taken of animals which may have been brought to the islands by man, whether intentionally or by accident, in post-Columbian or prehistoric time. Some of these have evolved under insular conditions into races distinct enough to be recorded as species or subspecies. The majority are identical with species of North or South America, Europe, Africa, et cetera. Some are known to have been introduced; others may be so explained by reason of associations of one kind or another. The formation of distinct races, such as are classed as species by modern mammalogists, does not necessarily take many centuries under these conditions—as witness the Porto Santo (Madeira) rabbits and other instances. Some of these supposedly introduced forms may have been brought in through natural means and be truly indigenous, although not very ancient; but it is impossible to prove such cases. It seems better here to omit all this doubtful evidence and consider only the fauna that is proved to be indigenous either by occurrence in the Pleistocene cave and spring deposits, by its sharp distinctions from any continental relatives, or by the high improbability that the animal could have been transported through human agency. Dr. G. M. Allen has com-

or possibly three of the numerous hystricomorph rodent stocks of Miocene South America, while the edentates represent only one group of the ground-sloths, the three or four other ground-sloth groups, as well as the several kinds of armadillos and the glyptodonts, being quite unrepresented.

The Insectivora

It appears to be reasonably certain that the Antillean rodents and edentates came from South America and from Tertiary South America. Hystricomorph rodents and edentates are unquestionably South American Tertiary types, which invaded North America when the two continents were joined, towards the end of the Tertiary. The insectivores, however, are more probably derivable from North American sources; for, with a single somewhat doubtful exception, the entire order of Insectivora is absent from the Tertiary faunas of South America, while they were many and varied in North America, especially in the older Tertiary. The two Antillean insectivores are not nearly related to each other, nor to any other genera of the order; they are placed in families by themselves. It has been repeatedly stated that *Solenodon* is related to the Malagasy Centetidæ, but in fact the affinity is a very distant one. *Nesophontes* is equally peculiar, and while it has some affinities with the Soricoidea (moles and shrews), they are very distant. The nearest relatives of both—collateral ancestors, perhaps—are certain imperfectly known insectivores of very primitive type in the North American Eocene and Oligocene.

The Rodentia

The rodents are clearly of South American affinities. They are all hystricomorphs—a group chiefly South American since the Middle Tertiary (if not before). The only North American hystricomorph is the porcupine, Pleistocene and Recent, whose ancestors have been recognized in the South American Tertiary. No traces of the hystricomorphs have been discovered in the Tertiary of North America.[3] There are certain Old World hystricomorphs, the Hystricidæ and certain Octodontidæ, and

piled an annotated list of the West Indian mammals which includes both introduced and indigenous types, and Dr. Thomas Barbour has done the same for the Reptilia and Batrachia. The evidence therein summarized will be discussed in a memoir on Cuban fossil mammals now in preparation. [This memoir was to have been a joint contribution by several authors. Dr. Matthew's part, dealing with the edentates, was completed before his death, but as yet has not been published.—E.H.C.]

3. Except Leidy's *Hystrix* (*Hystricops*) *venusta*, based on two teeth of doubtful affinities and uncertain geologic age.

the early Tertiary Theridomyidæ of Europe have been considered ancestral to the group, but they have no significant relations to the Antillean genera and their affinities are disputed, so that they may be passed over for the present problem.

The Antillean hystricomorphs are clearly related to the South American types, but it is equally clear that the relationship is not close. There are apparently three groups. One, including *Amblyrhiza* of Anguilla, *Elasmodontomys* and *Heptaxodon* of Porto Rico, is related to the chinchillas, but not closely related. Anthony[4] places them in separate subfamilies. Miller[5] states that they are more nearly related to the extinct *Megamys* and its allies than to the living chinchillids. These genera (*Megamys, Tetrastylus,* et cetera) are found in the Entrerian, Rio Negran, Hermosan, and Araucanian formations of Argentina accompanied by a fauna which is closely related to the Pampean, but contains a few little altered survivals from the Santa Cruz Miocene and comparatively few of the North American invading types. All should, in my judgment, be referred to the Upper Pliocene; they are certainly much later than the Santa Cruz.

Although *Megamys* and *Tetrastylus* are without doubt more nearly related than any modern type to the Antillean chinchillids and are considerably older, they cannot be regarded as ancestral. The common ancestral stock is to be found in the Santa Cruz chinchillids, all of which are of small or medium size. These Santa Cruz species are much more primitive, and the precise relationships will require more careful study.[6]

The remaining Antillean rodents are of South American type and broadly derivable from Santa Cruz rodents, but their more exact affinities are disputed and require more thorough and critical consideration and, if possible, more complete material. A thorough revision of the fossil rodent faunas of South America is an almost necessary groundwork for a correct estimate of their affinities. It is clear, however, that with the exception of *Capromys* they are not very closely related to the South American rodents, the common ancestral stock dating back probably to Pliocene

---

4. H. E. Anthony: New fossil rodents from Porto Rico. Bull. Am. Mus. Nat. Hist. **37**: 185-186. 1917.

5. Miller: Bones of mammals from Indian sites in Cuba and Santo Domingo. Smithsonian Miscell. Coll. **66(12)**: 3. 1916.

6. Miller apparently associates the Santa Cruz fauna of southern Patagonia with the very different and much later Entrerian fauna of northern Argentina, as he speaks of the two as though they were essentially one fauna, and refers to the "enormous extinct Patagonian rodents" as the nearest relatives of the Antillean chinchillids. The largest Santa Cruz rodents are species of *Perimys,* which is not nearly related to the Antillean genera. Most of the species are quite small. As between the Santa Cruz and the Hermosan (Pliocene) group of faunas, the tendency to rapid increase in size and specialization of numerous phyla is very marked, and is further emphasized in the Pampean (Pleistocene) group of faunas.

or late Miocene, as Miller believes. *Capromys* (and *Geocapromys*) would seem to be an exception, being quite close to the Venezuelan *Procapromys.*

The Edentata

The edentates include four quite distinct genera from Cuba and a fifth from Porto Rico, all referred to the family Megalonychidæ, but not closely related to any of the mainland forms. The largest, *Megalocnus,*[7] is about the size of a black bear; the smallest, *Microcnus,* about the size of a cat, and there are two of intermediate size, *Mesocnus,* with a rather long, narrow muzzle, and *Miocnus,* with a broad, square muzzle. The Porto Rican genus is related to *Miocnus,* both having heavy triangular tusks like the modern *Cholœpus;* the other three genera have large tusks, but of a peculiar dished shape, with a tendency to approach towards each other like the incisors of rodents. (They are not at all of the scalpriform gnawing type, however.) While these ground-sloths are sufficiently related to the North American genus *Megalonyx* to be placed in the same subfamily, they are quite evidently not descended from it, but contemporaneous specializations from the primitive Megalonychidæ of the South American Miocene, as represented in the Santa Cruz fauna. Among the known genera of this fauna there is only one, *Eucholœops* (including *Megalonychotherium*), which can be regarded as ancestral either to the Cuban ground-sloths or to *Megalonyx.* The others all have the caniniform teeth much reduced or vestigial and in series with the cheek teeth. To the best of my judgment, the anatomical evidence is not decisive as to whether the five Antillean genera are descended from one or from two or more nearly allied Upper Miocene or Lower Pliocene genera, but leads to the conclusion that the common ancestor or ancestors was very close to or identical with the Upper Miocene ancestor of *Megalonyx,* and was either the genus *Eucholœops* or one or more genera closely allied thereto. The Antillean genera represent, therefore, only the megalonychine division of the Megalonychidæ. The other families of ground-sloths—Mylodontidæ, Scelidotheridæ, and Megatheriidæ—are not found; nor are there any armadillos, glyptodonts, or anteaters.

Bats and Birds

In addition to the terrestrial mammals, bats are numerous in the cave deposits, and a number of birds, lizards, crocodiles, and turtles have been found at the Ciego Montero locality and elsewhere.

---

7. *Megalocnus,* Leidy, 1868. Proc. Acad. Nat. Sci. Phila. **1868**: 179.

Concerning the bats, there is very little to say. Most of them are nearly allied to or identical with species now living on the islands. The Antillean bats include a number of peculiar genera, besides others common to the continental parts of tropical America. Their relations are much the same as those of the birds, and in either case it is obvious that the intervening seas would act as a hindrance to migration, but not as an absolute barrier, and would be more of a hindrance in some groups than in others. The result would be the presence of a number of peculiar types, preserved by relative isolation and specialized in adaptation to the peculiarities of their habitat, along with other widely ranging forms closely allied to or identical with those of the mainland. The distribution of the birds has been carefully studied by Chapman and others.

Reptiles

The distribution of the lizards has been recently studied by Dr. Thomas Barbour, and his conclusions as to the paleogeography are sharply at variance with mine, owing to different methods of interpreting the data. I shall not take this part of the problem up at present.

The fossil crocodiles have been examined by Dr. Barbour, who informs me that they are all referable to *Crocodilus rhombifer,* a species still living on the island of Cuba.[8] The origin of this species might be either North or South American; but too little is known of the phylogeny and distribution of the Tertiary Crocodilia for any conclusions to be drawn as to the time or method of its arrival.

The fossil chelonians have not been carefully studied, but they include two species—one a giant tortoise, *Testudo cubensis* Leidy, which, like the giant tortoises of the Galapagos and Indian Ocean islands, has the carapace much thinned out, so that the plates are apparently more or less discontinuous. There is one North American Pliocene species, *T. pertenuis* Cope, from Texas which has a remarkably thin carapace, but apparently not discontinuous. The precise significance of this species in the paleogeographic problem must also await more careful study. The genus *Testudo* occurs sparingly in South America, and is recorded as a fossil in the Pliocene and Pleistocene formations—not earlier, so far as I know. On the other hand, species of *Testudo* are the most abundant of fossils in

8. Leidy's *Crocodilus pristinus* (1868, *l. c.*) was based upon a vertebra not distinguished from *C. rhombifer.* The skulls obtained by La Torre and Brown represent, in Doctor Barbour's opinion, a series of growth stages of the modern species, the largest much exceeding any modern specimens. Part of a skeleton associated with one of the largest skulls equals or exceeds Leidy's type of *pristinus* in size.

the Oligocene to Pliocene formations of North America; in the Pleistocene and Recent their range is restricted to the Southern States and Mexico. The indications point, therefore, preferably to North American origin for this Cuban tortoise, although not decisively.

The second fossil chelonian is one of the Emydidæ, or marsh-turtles; it appears to be a species of *Graphemys,* probably the same as the still existing Cuban species, which is said to be a close ally of *G. scripta* of the Southeastern States. Whether the two are specifically distinct has been questioned. The discovery of this species (if it be the same) fossil in the Ciego Montero locality removes any doubt as to its being indigenous to the island, as it carries it back into the Pleistocene, and probably to a time before the arrival of man. Its close relationship with *G. scripta* and limitation to the western islands, Cuba and Haiti, is a strong indication of its having come from Florida, and the time of its arrival probably would be not earlier than Pleistocene or at most late Pliocene.

SUMMARY OF AFFINITIES AND PROBABLE ORIGIN OF THE VERTEBRATE GROUPS

Summing up the indicated sources of the fossil and recent vertebrate fauna, we find it to be as follows:

1. The Insectivora, *Solenodon* and *Nesophontes,* of very ancient arrival, probably early Tertiary, and apparently of North American origin.
2. The ground-sloths, *Megalocnus, Mesocnus, Miocnus, Microcnus* in Cuba and *Acratocnus* in Porto Rico, of moderately ancient arrival, probably late Miocene or early Pliocene, and of undoubted South American origin. The rodents, with the exception of *Capromys,* fall also into this category.
3. The peculiar groups of birds, bats, and lizards are also no doubt of comparatively ancient arrival, but their source is unknown, as we know nothing of the Tertiary distribution of related groups on the mainland. The modern distribution of such related groups cannot be relied on, for we know that among terrestrial mammals various groups which are today exclusively or chiefly Neotropical were Nearctic until the end of the Tertiary and unknown in South American faunas until the late Pliocene. Presumably corresponding changes in distribution have occurred among the bats, birds, and reptiles, but we have no records as to what groups were affected. The Cuban crocodile may also be placed in this category.
4. Of the two chelonians the giant tortoise may be placed as more probably of North American than of Central or South American origin, but its time of arrival can hardly be estimated until its relations to the continental Tertiary species are known. The terrapin is almost certainly of North American origin, derived from the Southeastern States, and of comparatively late arrival, probably late Pliocene or Pleistocene.
5. *Capromys* and *Geocapromys* are undoubtedly of South American derivation, like the other rodents; but, so far as may be judged from their comparatively near affinity with the Venezuelan *Procapromys,* are of later arrival, perhaps late Pliocene or Pleistocene.

It appears, therefore, in sum, that the vertebrate fauna, fossil and recent, represents only a few selections from the continental faunas of either North or South America; that it falls into several groups of diverse origin, and judging from their degree of differentiation, of diverse times of arrival.

IS THE INCOMPLETE AND UNBALANCED CHARACTER OF THE FAUNA REAL OR ONLY APPARENT?

In the absence of hoofed animals, which form the greater part of all continental faunæ, in the tendency of races normally of small size to assume relatively large size and importance, in the relative fragility, so to speak, of the fauna, leading to its early disappearance when man invades the region—in many further points of detail—it parallels the faunas of those islands which lie beyond the continental shelf, and differs from those islands which lie within the shelf. In particular, the parallelism with the Madagascar fauna is made much closer by the recent discoveries. On the other hand, the contrast with the fauna of continental islands such as Borneo or Sumatra is a marked one. On these islands the fauna, although considerably specialized by isolation in Borneo, less so in Sumatra, is a fairly representative one. It includes all or nearly all of the important mammalian groups of the mainland save those which there is reason to believe are of too recent arrival or of unsuitable habitat to be present.

It may be objected that this difference is merely apparent; that the Pleistocene fauna of the Antilles was really of continental character, but because they are islands and not continents, it has been easily exterminated by man, and that the cave and spring deposits present only two distinct and very limited facies, not including the ungulates, carnivores, et cetera, which have been present. The best reply to this objection is to test it by comparison. Sumatra or Java are islands of comparable size to Cuba; Borneo is larger; Formosa or Hainan are comparable to Porto Rico. In none of these islands has the indigenous fauna been wiped out to anything like the extent necessary to obliterate its continental character, although all have been inhabited by man for a much longer time and in much larger numbers than the Antilles. The indigenous faunas of Great Britain or Ireland are far from exterminated, in spite of the great density of population and of modern civilization.

Nor can we assume that a cave or a spring fauna is so limited in its facies as to disguise a continental fauna type. The faunas of numerous caves in Europe and North America have been examined, and wherever any considerable collection is obtained it is clearly representative of the continental type. Spring or bog faunas sometimes contain little except hoofed animals, but I never heard of one in which hoofed animals and Carnivora were absent. I cannot escape from the conclusion that the Pleistocene fauna of Cuba was not a normal fauna, but deficient in most of the more abundant groups and composed of a selection of a very limited number of types which has expanded to a disproportionate variety and importance, owing to the absence of the rest of the fauna.

CONCLUSIONS AS TO FORMER GEOGRAPHIC RELATIONS AND MANNER OF COLONIZATION

As to the diverse origin of the several groups and the varying time during which they have been isolated on the islands, I have stated my interpretation of the evidence. I do not feel, however, that evidence of this sort leads to positive and certain conclusions.

As to the former connection of the Antilles with each other and with the mainland, my conclusions with the proviso just stated are as follows:

1. That the Greater Antilles have probably been united with each other, as far east as the Anguilla bank, in the late Tertiary or Pleistocene. This I conclude from the near affinity of representative species of the same or closely allied genera and the general similarity of the fauna, so far as known in the different islands.

2. That they have not at any time during the Tertiary been united with North America. If they had been we should find North American ungulates, rodents, carnivores, et cetera, differentiated in accord with the length of subsequent isolation, but of clearly recognizable affinities, and it would be a balanced or representative fauna. We might object that such a fauna had perhaps existed, but been wiped out by subsequent submergence. But the presence of *Solenodon* and *Nesophontes* negatives that, for they represent a very ancient survival, and if there had been a representative fauna it is hardly credible that submergence would have spared just two insectivores and destroyed all the rest of the fauna.

3. That they probably have not been connected with South America, either via the Lesser Antilles or via Central America, during the Tertiary; for if they had the fauna should be continental of South American type, with South American ungulate groups, marsupial carnivores, and a full representation of the rodents, edentates, et cetera.

4. The mammalian fauna appears to me to be reducible to perhaps three primary rodent stocks, one or more primary ground-sloth stocks, and two Insectivora. These I conceive to have arrived at various times during the Tertiary, the rodents and ground-sloths from South or Central America, the insectivores from North America, by accidents of transportation, of which the most probable for the mammals would perhaps be the so-

called "natural rafts" or masses of vegetation dislodged from the banks of great rivers during floods and drifted out to sea. The probabilities of this method I have elsewhere discussed.[9] For birds and bats, for the smaller reptiles, amphibians, fishes, and invertebrates, the problem of oversea transportation is a much simpler one.[10] That successful colonization in this way can occur is shown by their presence on nearly all oceanic islands; for it will hardly be maintained by reasonable men that every oceanic island has been joined to the mainland and has been continuously above water since its separation. Obviously, the larger the island and the nearer to continental land, the more often such colonization will occur.

5. The geology of the Caribbean region appears to me to afford no positive evidence against union of the Antilles either with South America or Central America; but neither does it afford any evidence that there ever was such union. Undoubtedly there is a line of disturbance and uplift along the Lesser Antilles, and another stretching through Haiti and Jamaica to Nicaragua; but evidence of similar and contemporaneous upheavals and similar sedimentation in two portions of this line of disturbance that are now separated by abyssal depths does not in the least prove that the intervening depths were formerly continuous land bridges. They may have been, but I do not see how any geologic evidence can prove that they were so. If we have evidence from some other source that there must have been a land bridge somewhere, then these lines of disturbance show its most probable location. That is all.

Land union with Florida appears to be distinctly against the geologic evidence, as in this region we have extensive flat-lying Tertiary marine and littoral formations which indicate that there has been very slight movement during the Tertiary, and that the present limits of the continental shelf represent probably the extreme extension of the land in the Pleistocene. Dall has shown the evidence very clearly in the case of Florida. Apparently the conditions in Yucatan are partly similar, but Vaughan has shown that its tectonic relations to the Antillean ridges are more favorable to a former union.

---

9. W. D. Matthew: Climate and evolution. Annals N. Y. Acad. Sci. **24**: 206. 1915.

10. Tropical storms, as Wallace pointed out years ago, probably play a principal part in transportation of very small animals or their eggs. Mammals could hardly be carried that way nor survive if they were.

Reprinted From

Scientia

Volume XLVIII, pages 33-42

July 30, 1930

THE DISPERSAL OF LAND ANIMALS

By William Diller Matthew

The present distribution of land animals is conditioned partly by climate and environment, partly by the configuration of the land areas of the globe. Few animals are cosmopolitan; for the most part those of the tropics are different from those of temperate and arctic zones, those of desert and plain from those of forest and swamp, while the oceans and mountain barriers that separate the continental regions mark out a third line of division. On the basis of these faunal distinctions the world has been divided into zoölogical regions and sub-regions, each characterized by its peculiar fauna. The boundaries and status of these regions have been much disputed because the barriers that confine one class of animals are not equally or similarly effective with other classes, and the faunal divisions, associations and boundaries vary accordingly. The regions most generally accepted are based primarily upon the distribution of terrestrial mammals. They conform less exactly to the distribution of birds and of lower vertebrates, or of the various invertebrate groups. Broadly speaking, geography has played the most important part in mammalian distribution. Climate and environment become relatively more and more important as we go down the scale.

If we examine the configuration of the present land areas, not on the misleading Mercator projection but on a globe or hemisphere-maps, it will appear that the northern continents form essentially one great land mass, divided only by the narrow and shallow strait between Alaska and Siberia. From it project the three southern continents as isolated peninsulas, South America and Africa connected by narrow isthmuses, Australia isolated save for a chain of islands. Antarctica is wholly isolated, with a wide stretch of deep ocean all around it. Besides these ocean barriers, two other barriers have been of prime importance in limiting distribution. The great mountain barrier of the Himalayas separates southern Asia and the associated islands as a distinct region. The great desert barrier of the Sahara cuts off central and southern Africa from the northern littoral.

Another physiographic feature of prime importance is the continental shelf. The continents are essentially broad platforms, raised in some areas into mountains or plateaux but averaging only a few hundred feet above sea-level, and with some portions of the platform submerged a few hundred feet below sea-level. The borders of the platform are everywhere sharply defined by the continental shelf, and beyond them the ocean suddenly deepens to several thousand feet, with a comparatively level floor of an average depth of two miles.

Most of the islands of the ocean lie upon these submerged banks, within the continental shelf and connected with the mainland by water of less than two hundred metres depth. Of the islands that lie beyond the shelf a few are of large size, most of them small volcanic or coral islands. The continents and the larger oceanic islands are substantially in isostatic balance throughout. That is to say, they are underlain by lighter rocks than those beneath the oceans, and are essentially floating at their present sea-level. The smaller islands are uncompensated so far as known.

The writer believes that the continental platforms and ocean basins are for the most part of immense antiquity. Large parts of them have been submerged in shallow water, re-elevated, eroded, and their materials shifted about, with a general tendency to deposit sediments within or upon the margin of the shelf, while the soluble parts of the rocks, lime, magnesia, iron, are more uniformly distributed wherever life or chemical agencies may precipitate them from the ocean water. The epicontinental and littoral sediments, dominantly silica and potash-soda-alumina silicates, consolidate into the lighter acid rocks; the lime-magnesia-iron, dominant in the oceanic sediments, with a smaller increment of cosmic dust, accumulate far more slowly and consolidate into heavier basic rocks. Hence isostasy results in a general tendency to build out the margins of the continents into steep-fronted deltas, and to raise elevated strips and mountain chains along their borders; and on the other hand to deepen the ocean basins progressively and to differentiate progressively the «sial» platform from the «sima» deep.

The more stable land areas have been refuges and centres of dispersal for terrestrial life since the geologic record began, and their relationships have been one of the two varying factors of dispersal and distribution.

Modern geologic knowledge disproves the old concept of a progressively cooling earth with its attendant climatic phenomena through the successive geological periods. It indicates instead an alternation of (a) climatic conditions not unlike those that prevail today, with (b) periods of more uniform climate all over the earth. The zonar, widely varying climates were associated with uplifted mountain regions and general elevation of the continental platforms, the uniform climates with reduction of the mountains and plateaux to vast peneplains by prolonged erosion, and flooding of the low-lying portions of the continental platforms by the ocean. The contrasted temperature zones result in more active atmospheric circulation, and with the high mountain ranges result in sharp contrasts of rainfall in different regions, extensive desert areas contrasting with areas of heavy rainfall; the more uniform climates with low relief of continents would result in more uniformly distributed rainfall averaging high because of the smaller land areas and larger ocean area of evaporation. Periods of extreme elevation culminate in glacial epochs, when great ice caps cover considerable regions of the land. From such an epoch the world is to-day but half emerged. Greenland and Antarctica are still buried deep beneath the ice. Another great glacial period of at least equal severity, with no less elevation and extension of the continents, followed the Coal Period, and separated the Paleozoic from the Mesozoic era. A third, probably of no less intensity, preceded the Paleozoic era. Other epochs of uplift, sometimes accompanied by evidences of glaciation, precede or intervene between these three great landmarks of geologic time; but they are less extreme or less clearly proved to be widespread or universal. Some of them may be local, not involving a general alteration of climate or geography.

The alternate periods of uniform climate are marked by wide extension of limestone formations over the shallow continental seas, followed often by extensive coal formations accumulated in vast swamps under a climate still moist and mild.

These general conditions of climate and geography have been everywhere modified by local geologic movements, and it is only the great outstanding extremes that are generally recognized as universal. Nevertheless the correspondence in character and succession of formations in different parts of the globe indicates that to a great extent each of the accepted geologic periods represents a climatic cycle of this kind. It begins and closes with an epoch of uplift, and includes a far longer period of quiescence, with progressive flooding and subsequent slow emergence of the continental platforms.

The fossil record of marine life is most complete during the epochs of submergence, more broken and scanty during epochs of emergence. The reverse might seem to hold true for terrestrial life; but in fact the limited extent and great thickness and unfavorable conditions for preservation of organic remains in the formations of the emergent phases, together with the fact that so large a part of them were deposited at the extreme margin of the continental shelf and hence now beyond our reach, makes the record of terrestrial life relatively poor, even in this phase.

Every species of animal is an adaptation more or less perfected to the needs of its particular mode of life. It has to compete with rivals and maintain itself against enemies and adverse features of environment. If the environment remains constant the species will in time perfect itself along the line in which it is specializing. If the environment changes, whether the change be in the physical surroundings or in the associated fauna, the species must either adapt itself to the new environment, seek elsewhere for a continuance of the old, or become extinct. Evolution, migration and extinction are thus conditioned by a changing environment.

The periods of continental uplift and zonar climates have compelled a great amount of change in terrestrial faunas. Some animals adapted themselves to new and more varied environment and conditions of life. A great adaptive radiation took place. Others migrated to regions where their old environment still prevailed, and, competing with the autochthonic fauna, introduced new conditions of survival which induced profound evolutionary changes in both native and invading faunas. Others again, unable to adapt themselves successfully or to find a refuge, became extinct, and their disappearance from the fauna left gaps in the interrelationship which were filled by a further series of adaptive readjustments.

It is obvious that the first effects of the oncoming zonar climates would be felt in the regions nearer the poles. The fáuna must be adapted to the severe conditions initiated there and spreading gradually over the temperate regions; while the tropical forest regions would serve as a refuge for less progressive types. The partially adverse environment and severe competition in the northern regions will tend to speed up selection and progressive evolution, and in result the polar regions will be centres of dispersal of higher, more dominant types. Mountain regions play a nearly similar rôle, and plains and steppes one partly similar.

The extension of the continents to their utmost borders, and connection of the outlying ones with the central mass, provides a greater area for terrestrial evolution and tends toward the production of cosmopolitan faunas.

Hence, in the zonar phase of the climatic cycle we expect to see new and dominant types evolved in the northern land masses, and thence invading the southern continents, where they would meet with a corresponding group or groups of dominant types evolved in the south and pressing northward. If the general relations of the continents were in the main as they are to-day, the southern groups, while corresponding to those of the north, would be relatively backward and weak because evolved in smaller and isolated areas. If Professor Wegener's hypothesis of continental drift or the very widely accepted views of Gondwana Land were true, then the southern dominant types should be equal to or stronger than those of the north, and the history of dispersal ought to be along the lines so conditioned. Among mammals, for instance, the notoungulates, the marsupial carnivores, the edentates should have been the source of the existing mammals of the northern world, and the artiodactyls, perissodactyls, Carnivora, etc., should have been now extinct or surviving only as a few rare types in tropical forests.

The great mountain ridges and deserts of the Old World, lying athwart the lines of polar dispersal, have served as barriers more effective or less so according to circumstances. In the New World the mountain ranges and desert strips lie radial from the northern and southern centres, so that the faunas in both North and South America are more uniformly distributed as to latitude.

(It is to be observed that it is not the barrier that limits distribution in many instances, but the sharp diversity of climate and environment conditioned by or associated with that barrier. Eastward of the Himalayas the Oriental grades into the Palæarctic fauna conformant to the gradation of climate. The boundary between Nearctic and Neotropical faunas lies not at the long, winding, narrow Isthmus of Panama, but where the Mexican plateau breaks down into the Central American lowlands.)

The broad lines of dispersal of land animals are obviously in accord with the foregoing principles. Among mammals it will hardly be questioned that the highest, dominant types and groups, those of latest evolution, are found chiefly in cold or temperate climates, in the northern world, in mountain and in plain faunas; while primitive survivals are chiefly tropical or southern and forest-living. Among lower vertebrates and the various groups of land invertebrates the same general arrangement appears to hold true in each group. Advanced progressive types are either confined to the northern world or have become cosmopolitan. Primitive survivals are found in tropical forests or in southern continents or in islands, especially tropical or southern islands.

Turning to the fossil record of land animals we find the inferences from modern distribution confirmed whenever the record is adequate. It makes clear, however, that most of the present distribution of mammals is the result of dispersal and migration during the Tertiary period. To a certain extent this is superposed upon an older dispersal, apparently Mesozoic, but so imperfectly documented that the direct evidence is of no great value. To this older dispersal we owe probably the Monotremes of Australasia, and perhaps the first arrival of certain marsupials and placentals in the southern continents; but the adaptive radiation of the several groups of marsupials in the southern continents, of edentates and notoungulates in South America, of hyracoids in Africa, took place during the Tertiary period coordinately with the adaptive radiation of most of the placental orders of mammals in the northern world, and the latter progressively invaded the tropical and southern continents as these became connected with the northern land mass.

While the geologic record of the lower land vertebrates and invertebrates is far less complete, it conforms apparently to the same principles, but it would appear that a more considerable part of their present distribution is due to pre-Tertiary dispersals; although it would seem that even with these the Tertiary adaptation and dispersal has been responsible for all the details and much of the major group distribution. Our knowledge of pre-Tertiary land faunas is relatively scanty, and mostly limited to certain restricted facies, so that the problem is more obscure and very liable to misinterpretation, especially with somewhat superficial acquaintance with the data.

Turning from generalities to detailed examination of a few sample groups, we find that the distribution of man conforms to these principles. He is today cosmopolitan, but the dominant races are native to the northern world, have largely invaded the south temperate and tropical regions, and the tropical forests are most resistant to their colonization. In these tropical forests survive more primitive human stages, also the anthropoid apes, the Old World and New World monkeys, and in the more outlying regions of forest the still more primitive lemuroids. Monkeys occur in the later Tertiary of Holarctica, lemuroids in the older Tertiary, but it is not demonstrated by the evidence that Palæarctica was the centre of dispersal of the primates, although the data may be so interpreted.

The Tertiary record shows that the Canidæ, now cosmopolitan, evolved in North America, that the Felidæ, now cosmopolitan, evolved somewhere in Holarctica, that the Ursidæ and Mustelidæ were likewise of Holarctic origin, that the Procyonidæ, now chiefly Neotropical, came from North America, and the Viverridæ, now chiefly Oriental and African, evolved in Palæarctica, and that South America lacked true Carnivora until the Pliocene, Africa until the Miocene, Australia still lacks them; but each southern continent evolved in their place parallel groups of carnivorous marsupials, which became extinct when the true carnivores invaded that continent.

Among the artiodactyl ungulates the dominant modern group of ruminants is shown to have dispersed from Palæarctica; relatively primitive survivals formerly inhabiting Holarctica are the tragulines of the Oriental region, *Hyæmoschus* of the African forests, the giraffe and okapi. The camels originated in North America, spread to South America and Asia and North Africa, and have become extinct in their original home. The pigs originated in Palæarctica and spread to the Oriental and African region, the peccaries in North America and have invaded South America and are now Neotropical. The perissodactyl ungulates have a corresponding history. They originated in Holarctica, and the most primitive living group, the tapirs, are now Oriental and Neotropical, although formerly Holarctic; the rhinoceroses, formerly Holarctic, are now Ethiopian and Oriental. The horses, the most specialized, originated apparently in North America, became cosmopolitan, and still survive in Africa and Asia. One might go through the list of the various mammalian groups, finding, wherever the evidence is adequate, that the great majority, and nearly all the dominant modern types, are of northern origin.

On the other hand, certain groups of mammals are shown to have undergone a great adaptive radiation and expansion in the southern continents, although the evidence goes to show that the primary types were originally invaders from the north. The epoch of uplift at the end of the Cretaceous enabled them to reach the southern continents, and during the isolation of the earlier Tertiary they expanded into faunas of varied adaptations, only to be overwhelmed by later northern invasions. In South America the edentates, notoungulates, litopterns, and the marsupial carnivores seem to have been autochthonous since the end of the Cretaceous, and, except for a few edentates, have wholly disappeared; the hystricomorph rodents and platyrhine monkeys appear to have arrived about the middle of the Tertiary, and a large proportion of these still survive. In Africa the fossil record is less complete, but the hyracoids, arsinoitheres and perhaps the Proboscidea would seem to represent the autochthonous early Tertiary fauna, while a more continuous connection with Palæarctica gave rise to a succession of invasions of the northern faunas. Australia, on the other hand, more isolated, received no mammals except the marsupials, of very ancient arrival and adaptively radiated into a wide diversity of type, a few stray rodents and bats of much later arrival, and the dingo, a probable companion of the invasion by man. In absence of a great northern invasion, the marsupial radiation still survives in Australia.

New Zealand, still more remote was never reached by mammals, and certain birds partly took their place. Unfortunately we have no record of the Tertiary land fauna of Australasia and New Zealand, save for a single marsupial, *Wynyardia,* from the Tertiary of Tasmania. It is related to the modern phalangers.

Antarctica to-day affords no opportunity for the existence of a terrestrial fauna. In the Tertiary and in preceding periods of mild climate it probably was habitable. It is probable that it was heavily forested, with a flora related to the contemporary flora of the southerly continents and islands. If any land connection with South America, Australia or New Zealand existed, the flora would be closely related to that of the connected region; if not, it should be as diverse from any of them as they are from each other. A land connection again would involve a fauna largely identical, continued isolation, a fauna largely in common so far as those animals are concerned which could be transported by storms or other accidents, but not likely to include mammals. Birds would be present, and in absence of mammals would probably evolve terrestrial adaptations. Bats might do so, but this seems less probable.

I regard the existing penguins as probably a terrestrial bird readapted to their present life. It is of interest that the sole fossil evidence that we have of the Tertiary fauna of Antarctica consists of the Tertiary penguins of Seymour Island. Some of these are of gigantic size. A study of their adaptive differences from modern penguins might be illuminating.

The distribution of the lower vertebrates conforms in its broader lines to that of the mammals. None of them, however, are so much controlled by land connections. This is even more clearly the case with most invertebrates. Birds, bats, and insects can fly. Small animals or their eggs and spawn may be transported by storms or on masses of floating vegetation far more easily than mammals. While they may not normally surmount ocean barriers, yet their general presence on even the smallest and most remote of oceanic islands sufficiently proves that they do so occasionally. It is not within reason to suppose that *all* oceanic islands have had a mainland connection. Few of them appear to be of great geological antiquity, and their faunas are scanty and incomplete in proportion to their isolation and small area. The larger and older geologically, and the nearer to the mainland, the greater the chances of colonization and the longer the time for adaptive radiation within the island area, which is often indicated by the submarine topography as formerly larger than the present and more connected up. The distinction is clear between the strictly continental faunas of all islands within the continental shelf and the insular faunas, incomplete and peculiar in almost uniform accord with the relations above indicated. A few minor and partial exceptions (certain Mediterranean and East Indian islands) are in regions of exceptional disturbance and upheaval indicated both by the Tertiary geology of the land and the extreme irregularity of the sea bottom.

The principles above outlined are for the most part not new. They have been held by the more cautious and conservative geologists and zoologists for many years. They have always been opposed by others of a more radical and speculative trend of mind, to whom the fascination of daring and sensational conclusions is irresistible. In the face of this opposition the doctrine of the permanence of the continents and oceans has never been popular nor has the interpretation of distribution in terms of existing and well known agencies aided by the cumulative effects of time been generally approved.

There are, however, two principles of modern geology which were not known to the older writers, and which form an essential part of this interpretation. The first is that all the continents and probably all the older oceanic islands are in approximate isostatic balance (within a few hundred feet), but the small volcanic islands are not (and tend therefore to sink back when the force that uplifted them is spent). This affords a sound reason for the permanency of continental platforms and ocean basins, and strongly supports the arguments of Dana and others. The second principle is that of cyclic alternations of climate broadly corresponding to the geologic periods and eras, in place of the old concept of a gradually cooling earth. The effect of this is to modify the concepts of dispersal presented by Wallace and others, substituting a definite sequence of diverse successive phases of evolution and dispersal conditioned by the diverse phases of the climatic-diastrophic cycle, for the older concept of a continuously progressive change. It is apparent that most of modern distribution is a result of Tertiary dispersal, but that this is superposed upon one or more older cycles of dispersal.

REPRINTED FROM
UNIVERSITY OF CALIFORNIA SYLLABUS SERIES
SYLLABUS NO. 213, FIGURES 8-12
AUGUST, 1928

TABLES SHOWING THE PAST AND PRESENT ZOÖGEOGRAPHICAL DISTRIBUTION OF LAND MAMMALS

BY WILLIAM DILLER MATTHEW

TABLE 1

PAST AND PRESENT DISTRIBUTION OF MAMMALS IN HOLARCTIC REGION

H = groups of Holarctic origin
Ne = groups of Neotropical origin
E = groups of Ethiopian origin
? = groups of doubtful origin

		GEOLOGICAL RANGE IN HOLARCTIC REGION								
		Cretaceous	Paleocene	Eocene	Oligocene	Miocene	Pliocene	Pleistocene	Existing Fauna	
	MULTITUBERCULATA	—	—							
	MARSUPIALS									
H	DIDELPHIDAE (opossums)	—	—	—	—	···	···	—	—	Survive chiefly in Neotropical region.
	PRIMATES									
?H	HOMINIDAE (man)							—	—	Cosmopolitan. Tertiary history unknown.
?	SIMIIDAE (apes)			···	···	··· —	—	—		Old World only. Surviving in Oriental and Ethiopian regions.
H	CERCOPITHECIDAE (monkeys and baboons)			···	···	··· —	—	—	—	
H	LEMUROIDEA (lemurs)	···	···	—	—					Survive in Oriental and Ethiopian.
?	**CHIROPTERA** (bats)	···	··· —	—	—	—	—	—	—	
	INSECTIVORA									
H	MENOTYPHLA	···	···	—	—					Survive chiefly in Oriental and Ethiopian.
?	ZALAMBDODONTA	···	···	···	···	—				Survive in Ethiopian and West Indian.
?	ERINACEIDAE				—	—	—	—	—	Survive chiefly in Ethiopian and Oriental.
H	TALPIDAE (moles)			—	—	—	—	—	—	
H	SORICIDAE (shrews)			···	—	—	—	—	—	

TABLE 1 (*Continued*)

PAST AND PRESENT DISTRIBUTION OF MAMMALS IN HOLARCTIC REGION

H = groups of Holarctic origin
Ne = groups of Neotropical origin
E = groups of Ethiopian origin
? = groups of doubtful origin

GEOLOGICAL RANGE IN HOLARCTIC REGION

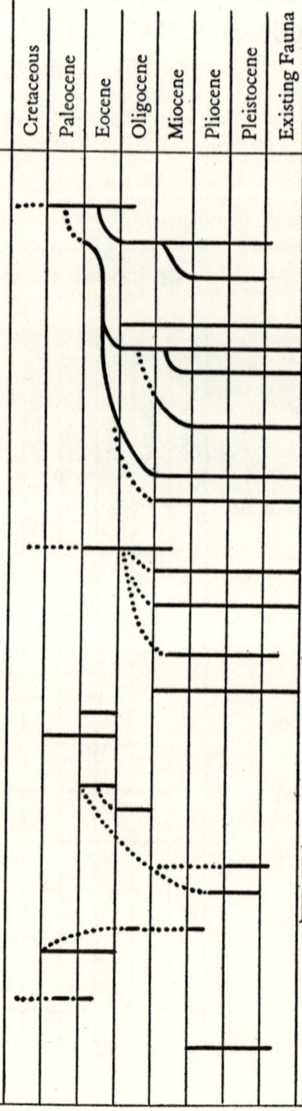

	CARNIVORA	
H	CREODONTA	
H	VIVERRIDAE (civets)	Old World only. Survive chiefly in Ethiopian and Oriental regions.
?	HYAENIDAE (hyaenas)	
H	FELIDAE (cats)	
H	CANIDAE (dogs)	
H	URSIDAE (bears)	
H	PROCYONIDAE (raccoons)	New World (except *Aelurus*). Survive chiefly in Neotropical region.
H	MUSTELIDAE (weasels)	
?	PINNIPEDIA (seals)	
	GLIRES (rodents)	
H	PROTROGOMORPHA	
H	SCIUROMORPHA	
H	MYOMORPHA	
?	HYSTRICOMORPHA	Survive chiefly in Neotropical region. Some in Africa and India.
H	LAGOMORPHA	
H	**TILLODONTIA**	
H	**TAENIODONTA**	
	EDENTATA	
H?	METACHEIROMYIDAE	
H?	MANIDAE (pangolin)	Surviving in Africa and India.
N?	LORICATA (armadillos)	Tertiary Neotropical. Reached North America in Pliocene.
N	GRAVIGRADA (ground sloths)	
H?	**TUBULIDENTATA**	Surviving in Africa.
H	**CONDYLARTHRA**	Probable source of many ungulate groups.
Ne	**NOTOUNGULATA**	Principal hoofed animals of Tertiary S. Am.
E	**HYRACOIDEA**	Mediterranean region. Chiefly African.

TABLE I (*Continued*)

PAST AND PRESENT DISTRIBUTION OF MAMMALS IN HOLARCTIC REGION

H = groups of Holarctic origin
Ne = groups of Neotropical origin
E = groups of Ethiopian origin
? = groups of doubtful origin

GEOLOGICAL RANGE IN HOLARCTIC REGION: Cretaceous, Paleocene, Eocene, Oligocene, Miocene, Pliocene, Pleistocene, Existing Fauna

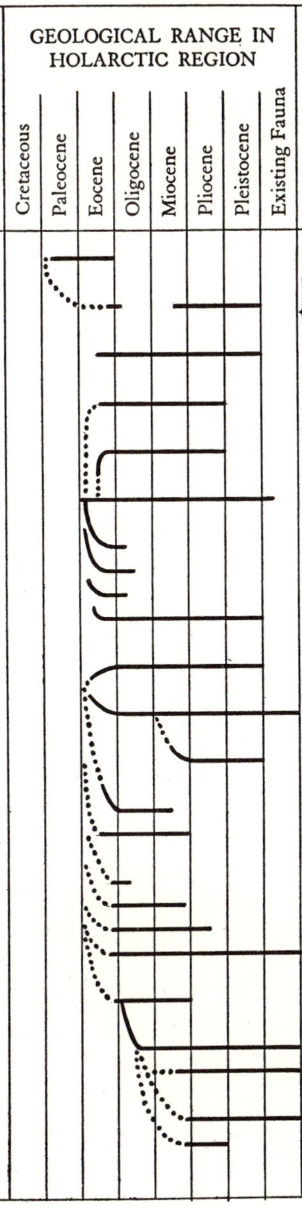

H	**AMBLYPODA**	
E?	**PROBOSCIDEA**	Early record in Egypt. Surviving in India and Africa.
?	**SIRENIA**	
	PERISSODACTYLA	
H	RHINOCEROTIDAE	Survive in Oriental and Ethiopian.
H	TAPIRIDAE	Survive in Oriental and Neotropical.
H	EQUIDAE	Survive in Asia and Africa.
H	LOPHIODONTIDAE	Chiefly European.
H	PALAEOTHERIIDAE	All European.
H	TITANOTHERIIDAE	
H	CHALICOTHERIIDAE	
	ARTIODACTYLA	
H	TAGASSUIDAE	New World only. Survive in S. America.
H	SUIDAE	Old World, now chiefly Oriental and Ethiopian.
?	HIPPOPOTAMIDAE	Old World only, surviving in Ethiopian.
H	ENTELODONTIDAE	
?	ANTHRACOTHERIIDAE	Survive to lower Pleistocene in India.
H	ANOPLOTHERIIDAE	Europe only.
H	CAENOTHERIIDAE	Europe only.
H	OREODONTIDAE	North America only.
H	CAMELIDAE	Survive in South America and Asia.
H	TRAGULIDAE	Survive in Oriental and Ethiopian regions.
H	CERVIDAE	
H	ANTILOCAPRIDAE	North America only. One surviving genus.
H	BOVIDAE	
H	GIRAFFIDAE	Survive to Pleistocene in India, recent in Africa.

TABLE 2

PAST AND PRESENT DISTRIBUTION OF MAMMALS IN NEOTROPICAL REGION

Ne = groups of Neotropical origin

H = groups of Holarctic origin

		CONTINENTAL SOUTH AMERICA						WEST INDIES		The record of Tertiary faunas in the Argentine Republic is more complete than anywhere else except Western Europe and the United States. Pleistocene fossils only are known from other parts of South America and the West Indies.
		Eocene	Oligocene	Miocene	Pliocene	Pleistocene	Modern Fauna	Pleistocene	Modern Fauna	
	MARSUPIALS									
H	DIDELPHIDAE	—	—	—	—	—	—			
Ne	BORHYAENIDAE	—	—	—	—	—				Take place of carnivora during Tertiary.
Ne	CAENOLESTIDAE	—	—	—	—	—	—			Take place of insectivora.
	PRIMATES									
?	CEBIDAE			—	…	—	—			Probably descended from Eocene lemuroidea of North America.
Ne	HAPALIDAE			…	…	…	—			
	CHIROPTERA	…	…	…	…	—	—	—	—	
	INSECTIVORA									
Ne	NECROLESTIDAE			—						Insectivora almost unknown in South America.
?	SOLENODONTIDAE								—	In West Indies only two peculiar and very primitive groups.
?	NESOPHONTIDAE									
H	SORICIDAE						—			
	CARNIVORA									Carnivora absent from South America until the later Pliocene. Their place in the fauna taken by carnivorous marsupials (Borhyænidæ).
H	FELIDAE					—	—			
H	CANIDAE					—	—			
H	URSIDAE					—	—			
H	PROCYONIDAE				—	—	—			
H	MUSTELIDAE					—	—			
	GLIRES									Hystricomorph rodents numerous and varied in South America and West Indies; others scarce or absent.
?	HYSTRICOMORPHA		—	—	—	—	—	—	—	
H	MYOMORPHA					—	—		—	
H	LAGOMORPHA						—			

TABLE 2 (*Continued*)

PAST AND PRESENT DISTRIBUTION OF MAMMALS IN NEOTROPICAL REGION

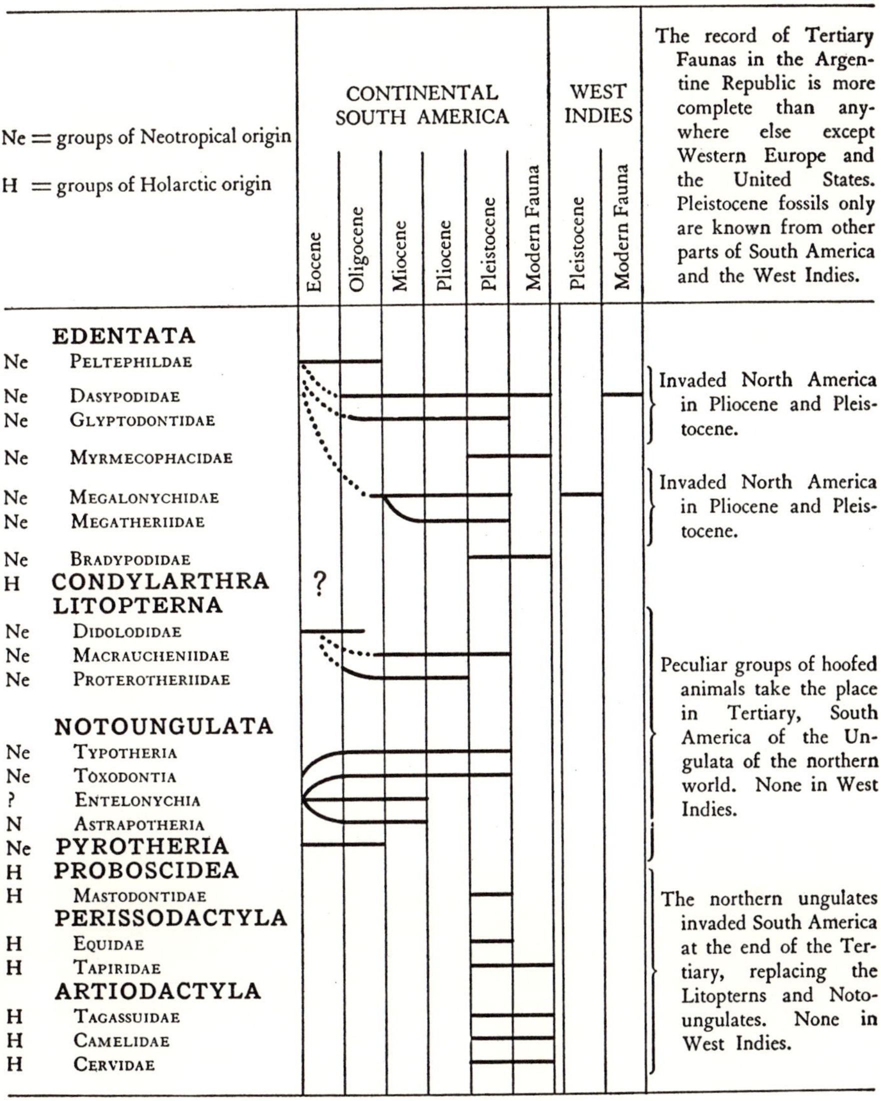

TABLE 3

PLEISTOCENE AND LIVING MAMMALS OF THE AUSTRALASIAN REGION

No Tertiary mammals have been discovered in Australia except a single specimen of a Diprotodont Marsupial *Wynyardia* from Tasmania.

	Pleistocene	Recent	Remarks
MONOTREMATA			
Ornithorhynchus	X	X	} Australia.
Echidna	X	X	
Proechidna	..	X	New Guinea.
MARSUPIALIA			
MYRMECOBIIDAE	..	X	Australian ant-eater.
PERAMELIDAE	X	X	Compare some rodents.
NOTORYCTIDAE	..	X	Mole-like.
DASYURIDAE	X	X	Take place of carnivora and rodents.
THYLACINIDAE—"wolf"	..	X	Take place of wolves.
PHALANGERIDAE—"cuscus"	X	X	Compare rodents.
Thylacoleo	X	..	
MACROPODIDAE—kangaroos	X	X	Take the place of most hoofed animals.
PHASCOLOMYIDAE—wombats	X	X	
DIPROTODONTIDAE	X	..	Compare rhinoceroses, etc.
CHIROPTERA			
PTEROPODIDAE—fruit bats	..	X	} Widely distributed in the tropics, some only Old World, some cosmopolitan.
EMBALLONURIDAE—large eared	..	X	
RHINOLOPHIDAE—nose leaf	..	X	
VESPERTILIONIDAE—"simple faced"	..	X	
CARNIVORA			
CANIDAE	X	X	The "dingo," perhaps introduced by man.
OTARIIDAE—eared seals	..	X	} Along coasts.
PHOCIDAE—earless seals	..	X	
RODENTIA			
MURIDAE	..	X	Six peculiar genera in New Guinea and Australia.

Various northern animals have been introduced by civilized man.

TABLE 4

PRESENT AND PAST OCCURRENCE OF MAMMALS IN THE ETHIOPIAN REGION

NOTE.—The past record is very imperfectly known. The Eocene and Oligocene faunas of Lower Egypt re probably partly of Ethiopian origin, although Egypt is not now in the Ethiopian region. Modern West frican faunas retain many primitive survivors. The modern Malagasy fauna is an expansion from a few nore or less primitive types. A few fossils have been found in the Miocene of east and South Africa.

= Holarctic origin = Ethiopian origin = Doubtful origin	EGYPT		EAST AND SOUTH AFRICAN MIOCENE			MODERN FAUNAS			Remarks
	Eocene	Oligocene	Miocene, East Africa	Miocene, South Africa	Pliocene, Egypt	Madagascar	West Africa	East and South Africa	
RIMATES									
? *Homo*	..	..	..	..	..	X	X	X	
? SIMIIDAE		?	..	..	..	..	X	X	
H CERCOPITHECIDAE ...	..	X	X	..	X	..	X	X	
LEMURIDAE	..	..	..	..	..	X	..	..	
H { INDRISIDAE	..	..	..	..	..	X	..	..	
H { LORISIDAE	..	..	..	..	..	..	?	..	
H { CHIROMYIDAE	..	..	..	..	..	X	..	..	
? GALAGIDAE	..	..	..	..	..	..	X	..	
NSECTIVORA									
H MACROSCELIDIDAE ...	..	..	..	..	..	..	..	X	Also Asiatic. Group Menotyphla.
H { CHRYSOCHLORIDAE ...	..	..	..	..	..	..	..	X	} Zalambdodonta.
H { POTAMOGALIDAE	..	..	..	..	..	..	X	..	} Zalambdodonta.
H { CENTETIDAE	..	..	..	..	..	X	..	..	} Zalambdodonta.
H TALPIDAE	..	..	..	..	..	..	..	X	
H SORICIDAE	..	..	..	..	..	..	..	X	
ARNIVORA									
CREODONTA	..	X	..	X	..	..	..	..	
VIVERRIDAE	..	..	..	..	..	X	X	X	
HYAENIDAE	..	..	..	..	..	..	..	X	
FELIDAE	..	..	..	..	X	..	X	X	
CANIDAE	..	..	..	..	X	..	..	X	
MUSTELIDAE	..	..	..	..	X	..	..	X	
PROTELIDAE	..	..	..	..	..	..	..	X	

TABLE 4 (*Continued*)

PRESENT AND PAST OCCURRENCE OF MAMMALS IN THE ETHIOPIAN REGION

H = Holarctic origin E = Ethiopian origin ? = Doubtful origin	EGYPT		EAST AND SOUTH AFRICAN MIOCENE			MODERN FAUNAS			Remarks
	Eocene	Oligocene	Miocene, East Africa	Miocene, South Africa	Pliocene, Egypt	Madagascar	West Africa	East and South Africa	
GLIRES									
H PROTROGOMORPHA	..	X	X	X	..	..	..	..	
H SCIUROMORPHA	..	..	..	X	..	..	..	..	
H MYOMORPHA	..	..	..	X	..	X	X	X	
? HYSTRICOMORPHA	..	..	..	X	X	..	..	X	
H LAGOMORPHA	..	..	..	..	..	..	..	..	
EDENTATA									
H MANIDAE	..	..	..	..	..	..	X	..	Pangolin.
? **TUBULIDENTATA**	..	..	..	..	..	..	..	X	Aardvark.
E **HYRACOIDEA**	..	X	X	X	..	..	..	X	Abundant and varied i Oligocene of Egypt.
E **ARSINOITHERIA**	X	X	..	..	..	..	..	..	
? **PROBOSCIDEA**									
DINOTHERIIDAE	..	..	X	..	X	..	..	..	
MASTODONTIDAE	..	..	X	..	X	..	..	..	
ELEPHANTIDAE	..	..	..	..	..	..	..	X	African elephant.
PERISSODACTYLA									
H RHINOCEROTIDAE	..	..	X	X	..	..	..	X	White and black Rhinoc- eroses. No tusks.
H EQUIDAE	..	..	..	..	X	..	..	X	Zebras in East and Sout Africa.
H CHALICOTHERIIDAE	..	..	X	..	..	..	..	..	
ARTIODACTYLA									
H SUIDAE	..	..	..	..	..	X	X	X	Modern bush pig i Madagascar.
? HIPPOPOTAMIDAE	..	..	X	X	X	X	X	X	Late Pleistocene in Ma- dagascar.
H ANTHRACOTHERIIDAE	..	X	..	..	..	..	..	..	
H TRAGULIDAE	..	..	..	..	..	..	X	..	*Hyaemoschus.*
H GIRAFFIDAE	..	..	..	..	X	..	X	X	Okapi in forests, Giraf in open.
H BOVIDAE	..	..	..	X	X	..	X	X	Abundant and varied.

TABLE 5

PAST AND PRESENT DISTRIBUTION OF MAMMALS IN ORIENTAL REGION

Families and Genera of Mammals	Eocene (Burma)	INDIA				Modern Oriental Fauna	Remarks
			Miocene	Pliocene	Pleistocene		
PRIMATES							
HOMINIDAE (man)					—	—	
SIMIIDAE (apes)				—	—	—	
CERCOPITHECIDAE			—	—	—	—	
LEMUROIDEA (lemurs)	···	···	···	···	···	—	Tarsier and Loris: probably indigenous through Tertiary.
INSECTIVORA							
TUPAIIDAE (tree-shrew)						—	Probably indigenous through Tertiary.
ERINACEIDAE						—	
SORICIDAE (shrews)						—	
CHIROPTERA	···	···	···	···	—	—	
CARNIVORA							
CREODONTA			—				Last known survivor of primitive carnivora.
VIVERRIDAE (civets)			—	—	—	—	
HYAENIDAE (hyaenas)			—	—	—	—	
FELIDAE (cats)			—	—	—	—	
CANIDAE (dogs)			—	—	—	—	Probably of North American origin.
URSIDAE (bears)			—	—	—	—	
MUSTELIDAE (weasels)			—	—	—	—	
GLIRES (rodents)							The fossil record of rodents is scanty, but squirrels, mice, and porcupines were probably common through later Tertiary.
SCIUROMORPHA						—	
MYOMORPHA				—	—	—	
HYSTRICOMORPHA				—	—	—	
EDENTATA							
MANIDAE (pangolins)						—	
PROBOSCIDEA							Proboscideans abundant and varied. The Oriental region may have been their chief center of evolution in later Tertiary.
DINOTHERIIDAE			—	—			
MASTODONTIDAE			—	—	—		
ELEPHANTIDAE			···	—	—	—	

TABLE 5 (*Continued*)

PAST AND PRESENT DISTRIBUTION OF MAMMALS IN ORIENTAL REGION

Families and Genera of Mammals	Eocene (Burma)		INDIA Miocene	INDIA Pliocene	INDIA Pleistocene	Modern Oriental Fauna	Remarks
PERISSODACTYLA							
AMYNODONTIDAE	—						
RHINOCEROTIDAE			—	—	—	—	Probably Palearctic origin.
TAPIRIDAE, etc.	—		·····	·····	—	—	
EQUIDAE (horses)				—	—		Of North American origin.
CHALICOTHERIIDAE			—	—	—		
TITANOTHERIIDAE	—						
ARTIODACTYLA							
SUIDAE (pigs)			—	—	—	—	
HIPPOPOTAMIDAE			·····	—	—		Survive in Ethiopian region.
ANTHRACOTHERIIDAE	—		—	—			Probably spread from Oriental region.
CAMELIDAE (camels)					—		North American origin. Reach Asia in Pliocene.
TRAGULIDAE	—		—	—	—	—	
CERVIDAE (deer)		···	—	—	—	—	Probably Palearctic origin.
GIRAFFIDAE (giraffes)			—	—	—		
BOVIDAE (antelopes and cattle)			—	—	—	—	Palearctic origin.

ANNOTATED BIBLIOGRAPHY OF WILLIAM DILLER MATTHEW[1]

By Charles Lewis Camp and Vertress Lawrence VanderHoof

1892a. On topaz from Japan. [Columbia College] School of Mines Quarterly **14** (**1**): 53-56. *4 figs.* 1892.
[Reprinted 1892 in, 'Contributions from the Mineralogical Laboratory of Columbia College **2**.']
[Technical descriptions of crystal faces, cleavages, and angles, of about 100 specimens.]

1893a. On phosphate nodules from the Cambrian of New Brunswick. Trans. N. Y. Acad. Sci. **12**: 108-120. *4 pls.* 10 Ap 1893.
[Reprinted 1893 in, 'Contributions from the Geological Department of Columbia College **9**.']
[Description of thin sections and hypotheses as to origin of the nodules. Sponge spicules and Foraminifera figured.]

1893b. On antennae and other appendages of *Triarthrus beckii*. Am. Jour. Sci. III. **46**: 121-125; and Trans. N. Y. Acad. Sci. **12** (**22**): 237-241. *pl. 7.* 17 Jl 1893.
[Reprinted 1893 in, 'Contributions from the Geological Department of Columbia College **14**.']
[One of the first descriptions of the head appendages of a trilobite. Agrees with Walcott that the group should be classed with the eurypterids and limulids, but "the characters seem to be of a more comprehensive type, approaching the general structure of the other crustacea rather than that of any special form."]

1893c. A study of the scale characters of the northeastern American species of *Cuscuta.* Bull. Torrey Bot. Club **20** (**8**): 310-314. *pls. 164, 165.* 10 Au 1893.
[Reprinted in, 'Contributions from the Herbarium of Columbia College **49**.']
[A key to the species, arranged according to scale characters, is given. Classification of the dodders is still one of the more difficult problems in systematic Botany. Use of scale characters is here for the first time extensively employed.]

1894a. The intrusive rocks near St. John, New Brunswick. Trans. N. Y. Acad. Sci. **13**: 185-203. *4 figs. pl. 5.* 1894.
[Reprinted in, 'Contributions from the Geological Department of Columbia College **22**.']
[Description and areal mapping of certain types of grano-diorite and olivine gabbro.]

1894b. The crystalline rocks near St. John, New Brunswick, Canada. Bull. Nat. Hist. Soc. N. B. **12**: 1-18. *map.* (Read May, 1894.)
"Includes sketch of the principles of metamorphism and igneous intrusions."

1895a. Monazite and orthoclase from South Lyme, Conn. Columbia College School of Mines Quarterly **16** (**3**): 231-233. Ap 1895.
[Specimens collected during a field trip in spring of 1894.]

1895b. The effusive and dyke rocks near St. John, New Brunswick. Trans. N. Y. Acad. Sci. **14**: 187-217. *2 figs. pls. 12-17.* 28 My 1895.
[Reprinted in, 'Contributions from the Geological Department of Columbia College **30**.']

---

1. Annotations in quotations from William Diller Matthew 1921n. Annotations in brackets by Camp and VanderHoof.

[Petrology, distribution and classification of these types of igneous rocks in the region. Submitted as Thesis for the Degree of Doctor of Philosophy in the Faculty of Pure Science, Columbia College, 1895.]

1895c. The volcanic rocks of the maritime provinces of Canada. Bull. Nat. Hist. Soc., N. B. **13**: 76-83. (Read May 7, 1895.)

[Outlines the volcanic history from Huronian to Triassic and discusses geologic history of the region.]

1896a. Metamorphism of Triassic coals at Egypt, North Carolina. (Abstract) Science **3**: 214. 1896.

1897a. Development of the foot in the *Palæosyopinæ*. Am. Naturalist **31**: 57-58. Ja 1897.

[Matthew's first paper in vertebrate paleontology. Distinguishes between broad and elongate footed types of Eocene titanotheres.]

1897b. Notes on intrusive rocks near St. John, New Brunswick, Canada. Bull. Nat. Hist. Soc., N. B. **15**: 61-64. (Read May 4, 1897.)

[Distribution and analysis of rock samples in continuation of a previous paper on the subject, 1894a.]

1897c. Status of the Puerco fauna. (Abstract) Anat. Anz. **14**: 231-232. 1897. [See also, Science **6**: 852.]

1897d. [Status of the Puerco fauna.] Trans. N. Y. Acad. Sci. **16**: 369-370. 1897.

[Notes on the "rodent" *Mixodectes*, later regarded as an insectivore, and statement that placental mammals become differentiated at beginning of Tertiary.]

1897e. A revision of the Puerco fauna. Bull. Am. Mus. Nat. Hist. **9**: 259-323. *20 figs.* 16 N 1897.

"Revision of the species and their affinities based on a restudy of the Cope collection along with American Museum collections made in 1892 and 1896. Torrejon and Puerco faunas distinguished. Cope's theory of primitive serial carpus and tarsus rejected."

1898a. [*Clænodon* and *Oxyæna* described.] (Abstract) Science **8**: 880. 1898.

1899a. A provisional classification of the freshwater Tertiary of the West. Bull. Am. Mus. Nat. Hist. **12**: 19-75. 31 Mr [-8 Ap] 1899.

"Revised list of mammal faunæ of American Tertiary formations with details of occurrence and critical estimate of the validity and importance of species. Correlation based upon these data."

1899b. Is the White River Tertiary an æolian formation? Am. Naturalist **33 (389)**: 403-408. My 1899.

[Yes, a loessic deposit.]

1899c. **Wortman, Jacob Lawson; & ——————**

The ancestry of certain members of the Canidæ, the Viverridæ and Procyonidæ. Bull. Am. Mus. Nat. Hist. **12**: 109-139. *10 figs. pl. 6.* 21 Je 1899.

"The sections by Matthew deal with the affinities of the Oligocene ('Miocene') Canidæ of North America and describe *Phlaocyon* as an ancestor of the Raccoons."

1900a. The Cope Pampean collection. Am. Mus. Jour. **1** (**2**): 24-26. *1 fig.* My 1900.

[This collection formed a part of the Argentine exhibit at the Paris Exposition in 1878. It was purchased by E. D. Cope and later by the American Museum.]

1901a. Additional observations on the Creodonta. Bull. Am. Mus. Nat. Hist. **14**: 1-38. *17 figs.* 31 Ja 1901.

"Classification of the Creodonta, with brief discussion of principles. New observations upon various Creodont genera."

[Favors a compromise between the "horizontal" and "vertical" systems of classification.]

1901b. Fossil mammals of the Tertiary of northeastern Colorado. Mem. Am. Mus. Nat. Hist. **1**: 353-447. *34 figs. pls. 37-39.* N 1901.

"Lacustrine origin of White River and Loup Fork groups rejected in favor of floodplain playa and æolian deposition. New theory of adaptation of sabre-tooth cats. Description of new collections obtained in 1898, and revision of various groups, Carnivora, camels, oreodonts, etc."

1902a. The hall of fossil vertebrates. Am. Mus. Jour. **2** (1) Supplement: 1-19. *11 figs.* Ja 1902.
[See 1903g.]

1902b. A skull of *Dinocyon* from the Miocene of Texas. Bull. Am. Mus. Nat. Hist. **16**: 129-136. *4 figs.* Ap 1902.
[A huge "bear-dog" not greatly resembling either bears or dogs.]

1902c. On the skull of *Bunælurus,* a musteline from the White River Oligocene. Bull. Am. Mus. Nat. Hist. **16** (12): 137-140. *3 figs.* 7 Ap 1902.
[A primitive member of the weasel-skunk group, approaching the cats in the reduction of the teeth.]

1902d. [Note on the exhibit, development of the horse.] Am. Mus. Jour. **2** (5): 40-41. My 1902. [Unsigned.]

1902e. New Canidæ from the Miocene of Colorado. Bull. Am. Mus. Nat. Hist. **16**: 281-290. *4 figs.* 18 S 1902.
"First outline of theory of Holarctic dispersal and climatic control (286-287)."
[Later elaborated in Matthew "Climate and Evolution" (1915d).]

1902f. A horned rodent from the Colorado Miocene. With a revision of the Mylagauli, beavers, and hares of the American Tertiary. Bull. Am. Mus. Nat. Hist. **16**: 291-310. *17 figs.* 25 S 1902.
"Description of *Ceratogaulus.* Affinities of Mylagaulidæ, *Steneofiber* and *Palæolagus.*"

1902g. The skull of *Hypisodus,* the smallest of the Artiodactyla, with a revision of the Hypertragulidæ. Bull. Am. Mus. Nat. Hist. **16**: 311-316. *4 figs.* 25 S 1902.

1902h. List of the Pleistocene fauna from Hay Springs, Nebraska. Bull. Am. Mus. Nat. Hist. **16**: 317-322. 25 S 1902.

1902i. Review. 'On Vertebrata of the Mid-Cretaceous of the North-West Territory. —(1) Distinctive characters of the Mid-Cretaceous fauna, by Henry Fairfield Osborn.'—(2) 'New genera and species from the Belly River series (Mid-Cretaceous), by Lawrence M. Lambe.' Ottawa Naturalist **16**: 169-170. N 1902.

1903a. Second Cope collection and Pampean collection. Am. Mus. Jour. **3** (1): 3-6. Ja 1903. [Unsigned.]
[About 4000 specimens, including a number of Cope's types.]

1903b. The evolution of the horse. Am. Mus. Jour. **3** (1) Supplement: 1-30. *4 figs. 7 pls.* Guide leaflet **9**. Ja 1903.
[Second edition May 1905; third edition 1909; fourth edition 1913b; fifth edition 1921a.]
"A popular guide to the American Museum collection of fossil Equidæ, pointing out the mechanics of the adaptive changes, and the environmental and climatic changes that brought them about."

1903c. News notes. Am. Mus. Jour. **3** (4): 58-61. 1903. [Unsigned.]
[Dept. of Vert. Paleont. 58-61; News notes 62-66.]

1903d. The fauna of the Titanotherium beds at Pipestone Springs, Montana. Bull. Am. Mus. Nat. Hist. **19**: 197-226. *19 figs.* 9 My 1903.
"Various new Lower Oligocene genera described and affinities discussed."

1903e. A fossil hedgehog from the American Oligocene. Bull. Am. Mus. Nat. Hist. **19**: 227-229. *1 fig.* 9 My 1903.
[The first true hedgehog to be found in America.]

1903f. Recent zoopaleontology (concerning the ancestry of the dogs). Science **18** (**440**): 912-913. 5 Je 1903.

1903g. The collection of fossil vertebrates. A guide leaflet to the exhibition halls of vertebrate palæontology. Am. Mus. Jour. **3** (5) Supplement: 1-32. *15 pls.* Guide Leaflet **12**. O 1903.
[A revision of 1902a.]

1904a. An extinct cave fauna in Arkansas. Am. Mus. Jour. **4** (1): 6-7. Ja 1904. [The Conard Fissure.]

1904b. Department of Vertebrate Palæontology: Field explorations in 1903. Am. Mus. Jour. **4** (1): 14-15. 1904. [Unsigned.]

1904c. Eocene fossil mammals of South America. Am. Mus. Jour. **4** (1): 26. 1904. [Unsigned.]
[Casts donated by Florentino Ameghino.]

1904d. A complete skeleton of *Merycodus*. Bull. Am. Mus. Nat. Hist. **20**: 101-129. *21 figs. pl. 3.* 31 Mr 1904.
"Related to *Antilocapra.*"

1904e. Notice of two new Oligocene camels. Bull. Am. Mus. Nat. Hist. **20**: 211-215. 2 Je 1904.
[*Pseudolabis* n. gen. and *Miolabis* n. subgen. and key to the genera of fossil camels.]

1904f. ——————; **& Gidley, James.**
New or little known mammals from the Miocene of South Dakota. American Museum Expedition of 1903. II. Carnivora and Rodentia by W. D. Matthew. Bull. Am. Mus. Nat. Hist. **20**: 246-265. *15 figs.* Jl 1904.
[Includes note on evolution of rodent teeth.]

1904g. Outlines of the continents in Tertiary times. Science **19**: 581-582. 1904.
[Also in Am. Geol. **33**: 268-269; treated more fully in Matthew 1906c.]

1904h. Exhibition of a series of foot-bones showing the evolution of the camel. (Abstract) Science **19** (**493**): 892. 10 Je 1904.

1904i. The arboreal ancestry of the Mammalia. Am. Naturalist **38** (**455-456**): 811-819. N-D 1904.
[Huxley and Dollo adduced arboreal ancestry for marsupials. Matthew extends the hypothesis to include placentals.]

1904j. Mode of evolution of the Vertebrata. The New International Encyclopædia. New York, 1904. 'Vertebrata.'
[Also appears in subsequent editions of this work.]

1905a. The fossil carnivores, marsupials, and small mammals in the American Museum of Natural History. Am. Mus. Jour. **5** (1): 22-59. *27 figs.* Ja 1905.
[Reprinted as Guide Leaflet **17**: 1-39.]

1905b. On Eocene Insectivora and on *Pantolestes* in particular. (Abstract) Science **21** (**530**): 298-299. 24 F 1905.

1905c. The mounted skeleton of *Brontosaurus* [in the American Museum of Natural History]. Am. Mus. Jour. **5** (2): 63-70. *4 figs.* Ap 1905.

1905d. Notice of two new genera of mammals from the Oligocene of South Dakota. Bull. Am. Mus. Nat. Hist. **21**: 21-26. *6 figs.* 14 F 1905.
[*Eutypomys* n. gen. of castorid rodents, and *Heteromeryx* n. gen. of Hypertragulidæ; with remarks on distinctions between White River hypertragulids.]

1905e. Notes on the osteology of *Sinopa,* a primitive member of the Hyænodontidæ. Proc. Am. Phil. Soc. **49**: 69-72. 1905.
[A more complete paper on this subject 1906a.]

1905f. [Outlines of the continents in Tertiary times.] (Abstract) Ann. N. Y. Acad. Sci. **16**: 315-316. 1905.
[See 1906c.]

1906a. The osteology of *Sinopa,* a creodont mammal of the Middle Eocene. Proc. U. S. Nat. Mus. **30**: 203-233. *20 figs. pl. 16.* 1906.
"Description of a complete skeleton in the National Museum; affinities of *Sinopa,* revision of the species."
[A primitive, cursorial creodont related to *Hyaenodon.*]

1906b. Field work of the Department of Vertebrate Palæontology. Am. Mus. Jour. **6** (**3**): 122-123. Jl 1906. [Unsigned.]

1906c. Hypothetical outlines of the continents in Tertiary times. Bull. Am. Mus. Nat. Hist. **22**: 353-383. *7 figs.* 25 O 1906.
"Contains a series of maps attempting to correlate views of various authorities on a conservative basis. The Antarctic connections shown on one of these maps were shortly afterwards abandoned as untenable."

1906d. Fossil Chrysochloridæ in North America. Science **24** (**634**): 786-788. 14 D 1906.
[A humerus from the Rosebud beds, Lower Miocene of South Dakota. Named *Arctoryctes terrenus* n. gen. and sp., without further description. See 1907c: 172.]

1907a. The skeleton of the Columbian mammoth. Am. Mus. Jour. **7** (**1**): 5-6. Ja 1907. [Unsigned and of doubtful authorship.]
[Discovered "four miles east of Jonesboro, Indiana," in a swamp deposit, referred to the Middle Pleistocene.]

1907b. Department of Vertebrate Palæontology. Field expeditions of 1906. Am. Mus. Jour. **7** (**1**): 6-8. *1 fig.* Ja 1907. [Unsigned.]

1907c. A Lower Miocene fauna from South Dakota. Bull. Am. Mus. Nat. Hist. **23**: 169-219. *26 figs.* 14 Mr 1907.
"First discovery of the true Lower Miocene fauna; discussion of correlation and affinities."

1907d. The *Naosaurus* or "ship-Lizard." Am. Mus. Jour. **7** (**3**): 36-41. *2 figs.* Mr 1907. [Unsigned.]

1907e. When brute force ruled the world. Discovery **1** (**2**): 42. My 1907.

1907f. The relationship of the 'Sparassodonta.' Geol. Mag. V. **4**: 531-535. D 1907.
[Marsupial relationships favored.]

1908a. Mammalian migrations between Europe and North America. Am. Jour. Sci. IV. **25**: 68-70. Ja 1908.
[Distribution of Tertiary faunas indicates that Asia "is and has been the great center of evolution and dispersion of dominant mammalian types." Basal Eocene, Middle Eocene and Middle Oligocene indicated as separation between North America and Asia.]

1908b. *Allosaurus,* a carnivorous dinosaur and its prey. Am. Mus. Jour. **8** (**1**): 2-5. *1 pl.* Ja 1908.

1908c. The new *Ichthyosaurus.* Am. Mus. Jour. **8** (**1**): 6-8. *1 pl.*
[Specimen from Holzmaden showing outline of body.] Ja 1908.

1908d. A four-horned Pelycosaurian from the Permian of Texas. Bull. Am. Mus. Nat. Hist. **24**: 183-185. *1 fig.* 13 F 1908.
[*Tetraceratops insignis* n. gen. and sp.]

1908e. Review. 'A revision of the Pelycosauria of North America' by E. C. Case. Science **27** (**699**): 816-818. 22 My 1908.

1908f. Osteology of *Blastomeryx* and phylogeny of the American Cervidæ. Bull. Am. Mus. Nat. Hist. **24**: 535-562. *15 figs.* 30 Je 1908.
"Description of complete skeleton, discussion of affinities. *Leptomeryx, Blastomeryx, Mazama, Cervus* regarded as successive structural (not genetic) evolutionary stages of the Cervidæ, derived from Palæarctic ancestry and successively invading the New World."

1908g. News notes. [The Trachodon group] Am. Mus. Jour. **8** (**6**): 89. O 1908. [Unsigned.]

1908h. News notes. [Additions to the hall of fossil mammals. Exhibition of *Orohippus osbornianus.*] Am. Mus. Jour. **8** (**6**): 89. O 1908. [Unsigned.]

1908i. Fossil fishes. Am. Mus. Jour. **8** (**7**): 110. N 1908.

1908j. Exhibit illustrating the evolution of the horse. Am. Mus. Jour. **8** (**8**): 116-122. *3 figs. 2 pls.* D 1908.

1909a. Observations upon the genus *Ancodon.* Bull. Am. Mus. Nat. Hist. **26**: 1-7. 5 Ja 1909.

[*Ancodon* (? *Bothriodon*) *leptodus* n. sp. from Lower Rosebud (Arikaree), Lower Miocene of South Dakota.]

1909b. Seventh annual meeting of the American Society of Vertebrate Paleontologists. Science **29** (**735**): 194-198. 29 Ja 1909.

[Signed by Matthew as 'Secretary.' Includes abstracts of papers by Matthew (On a skull of *Apternodus* and the skeleton of a new artiodactyl from the Lower Oligocene of Wyoming); by Matthew and Cook (Pliocene Fauna of Western Nebraska); and by Matthew (On *Bison latifrons*). For the complete paper on *Apternodus* see 1910d.]

1909c. Constitution and By-Laws of the Paleontological Society. 15 Mr 1909.

[Signed by John M. Clarke, T. W. Stanton, David White, C. R. Eastman, W. D. Matthew and H. F. Cleland. Reprinted, without signatures, in Bull. Geol. Soc. Am. **21**: 77-82. 1910. Matthew was elected treasurer to the Society at this time and later became its President.]

1909d. Recent purchases of fossil vertebrates. Am. Mus. Jour. **9** (**3**): 68-69. Mr 1909.

[The Sternberg *Trachodon* "mummy," skeleton of the marine turtle *Toxochelys* and skull of the mosasaur *Clidastes*.]

1909e. The oldest land reptiles of North America. Am. Mus. Jour. **9** (**4**): 91-95. *5 figs.* Ap 1909.

[Notice of the Cope Permian collection.]

1909f. **Osborn, Henry Fairfield; & ————**

Geological correlation through vertebrate paleontology by international coöperation. Ann. N. Y. Acad. Sci. **19**: 41-44. 20 Ap 1909.

[First report of Section of Vertebrate Paleontology of International Correlation Committee, National Academy of Sciences. Matthew was the Secretary of this Section. (Abstract) Ann. N. Y. Acad. of Sci. **19**: 302-303.]

1909g. **Osborn, Henry Fairfield; & ————**

Explorations and researches of the Department of Vertebrate Palæontology. Preface to: 'Fossil Vertebrates in the American Museum of Natural History, **3**: iii-xvi. 1904-1908.' Ap 1909.

1909h. 'The fossil vertebrates of Belgium' by Louis Dollo [translated by W. D. Matthew]. Ann. N. Y. Acad. Sci. **19**: 99-119. *pls. 4-10.* 31 Jl 1909.

1909i. The Carnivora and Insectivora of the Bridger Basin, Middle Eocene. Mem. Am. Mus. Nat. Hist. **9** (**6**): 289-567. *118 figs. pls. 42-52.* Au 1909.

"Review of the Bridger formation, the source and method of deposition of the sediments, stratigraphic and faunal divisions; adaptation and ecology of the fauna. Detailed osteological descriptions of Bridger Carnivora and Insectivora, discussion of affinities and classification; various new or little known types fully described."

1909j. **————; & Cook, Harold James.**

A Pliocene fauna from western Nebraska. Bull. Am. Mus. Nat. Hist. **26**: 361-414. *27 figs.* 3 S 1909.

"First description of the Snake Creek fauna, regarded as Lower Pliocene."

1909k. Faunal lists of the Tertiary Mammalia of the west. In 'Cenozoic Mammal horizons of Western North America.' By H. F. Osborn. U. S. Geol. Surv. Bull. **361**: 91-120. 1909.

"Lists revised down to 1908, as a basis for a more exact correlation." [The copy in the Matthew Library at the University of California contains Matthew's own manuscript annotations and emendations.]

1909l. Science-History of the universe. Zoology. New York. Current Literature Publishing Co. **6**: 1-194. 1909.

[Contains the following prediction, page 194: "In the present writer's opinion the geological evidence of the ancestry of man and the most direct phylogeny of many mammals will be discovered when the Tertiary formations of Central and Eastern Asia are adequately and thoroughly searched for fossil vertebrates . . ."]

1910a. Patagonia and the Pampas Cenozoic of South America. A critical review of the correlations of Santiago Roth, 1908. Ann. N. Y. Sci. **19**: 149-160. *pl. 14.* 1910.

[Includes list of "Characteristic genera of the South American Cenozoic with provisional ordinal references."]

1910b. The *Tyrannosaurus.* Am. Mus. Jour. **10** (1): 2-8. *6 figs.* Ja 1910. [Unsigned.]

[Describes the discovery and collecting of a complete skeleton, later mounted at the American Museum.]

1910c. A complete pterodactyl skeleton. Am. Mus. Jour. **10** (2): 49-50. *1 fig.* F 1910. [Unsigned.]

[*Pterodactylus elegans* from the Solenhofen beds of Bavaria.]

1910d. On the skull of *Apternodus* and the skeleton of a new artiodactyl. Bull. Am. Mus. Nat. Hist. **28**: 33-42. *5 figs. pl. 6.* 22 Mr 1910.

1910e. On the osteology and relationships of *Paramys,* and the affinities of the Ischyromyidæ. Bull. Am. Mus. Nat. Hist. **28**: 43-72. *19 figs. pls. 1-5.* 22 Mr 1910.

"Complete skeletons of Eocene rodent *Paramys* and skulls, etc. of other Eocene and Oligocene Rodentia described. The Eocene Ischyromyidæ regarded as the primitive group from which the other rodent groups, all of later age, have been diversely specialized."

1910f. The pose of sauropodous dinosaurs. Am. Naturalist **44** (525): 547-560. S 1910.

[Regards the giant sauropods as wading forms with upright limbs.]

1910g. The phylogeny of the Felidæ. Bull. Am. Mus. Nat. Hist. **28**: 289-316. *15 figs.* 19 O 1910.

"The Tertiary Felidæ regarded as representing two phyla, one leading from *Dinictis* up into the true cats, the other from *Hoplophoneus* into the sabre-tooth tigers. Nothing is known of their Eocene ancestry. Discussion of the adaptation of the sabre-tooths as shown by characters of skull and skeleton."

1910h. **Osborn, Henry Fairfield; Gregory, William King; Mosenthel, Johanna Kroeber; & ————**

Outline classification of the Mammalia recent and extinct. In 'The Age of Mammals.' By H. F. Osborn. New York: Macmillan. 511-563. 1910.

["This classification has been prepared under the direction of the author by William King Gregory and Johanna Kroeber Mosenthal. The Geological range and revision of the extinct genera has been done with the coöperation of William Diller Matthew.—H. F. O."]

1910i. Notes and literature—Schlosser on Fayûm mammals. Am. Naturalist **44** (**1910**): 700-703. 1910.

"A preliminary notice of Dr. Schlosser's studies upon the collections made in the Oligocene of Egypt for the Stuttgart Museum."

[Abstract in Sci. Progress **5**: 662-663.]

1910j. Continuity of development. Popular Science Monthly **77**: 473-478. N 1910.

"Contribution to a symposium of the Palæontological Society, dealing with the question of saltation versus continuous progression in phylogeny of fossil vertebrates. The conclusion is reached that the apparently sudden appearance of new types is usually due to migration or to gaps in the geologic record, the best known phyla progressing through very small changes. This is in no wise held to exclude the Mendelian character of such differences, but they are not normally large in amount."

1910k. The new Plesiosaur. A great marine reptile of the ancient world. Am. Mus. Jour. **10** (**8**): 246-250. *2 figs. 2 pls.* D 1910.

[*Cryptoclidus.*]

1911a. Fort Lee dinosaur. Am. Mus. Jour. **11** (1): 28-29. Ja 1911.

[Mentions discovery of remains later determined to be of the phytosaur *Clepsysaurus manhattanensis.*]

1911b. The ground sloth group. Am. Mus. Jour. **11** (**4**): 113-119. *2 figs. 2 pls.* Ap 1911.
[Figures experimental poses in making the group. Reprinted with additions by the Am. Mus. in 1913.]

1911c. A tree climbing ruminant. Am. Mus. Jour. **11** (**5**): 162-163. *1 fig.* My 1911.
[*Agriochoerus*, an oreodont.]

1911d. The amphibians of the great coal swamps. Am. Mus. Jour. **11** (**6**): 197-200. *2 figs.* O 1911.

1911e. Fossil vertebrates—what they teach. Am. Mus. Jour. **11** (**7**): 246-247. N 1911.
[Synopsis of three halls of fossil vertebrates in the American Museum.]

1911f. News notes. [Eocene fossil mammals.] Am. Mus. Jour. **11** (**8**): 311. 1911. [Unsigned.]

1912a. News notes. [New exhibits of fossil vertebrates.] Am. Mus. Jour. **12** (**1**): 37. Ja 1912. [Unsigned.]

1912b. News notes. [Prjevalsky wild horses.] Am. Mus. Jour. **12** (**2**): 76. F 1912. [Unsigned.]

1912c. Review of Abel's 'Grundzüge der Palæobiologie der Wirbelthiere.' Science **35** (**896**): 341-342. 1 Mr 1912.

1912d. Florentino Ameghino. Pop. Sci. Monthly **80** (**3**): 303-307. *3 figs.* Mr 1912.

1912e. The new four-toed horse skeleton. Am. Mus. Jour. **12** (**5**): 186. *1 fig.* My 1912.

1912f. African mammals. Bull. Geol. Soc. Am. **23**: 85, 156-162. 1 Je 1912.
[Discussion of the Fayûm and other fossil faunas. Origin and migrations of mammals.]

1912g. [Ten years progress in vertebrate paleontology:] Carnivora and Rodentia. Bull. Geol. Soc. Am. **23**: 85, 181-187. 1912.
"Critical reviews of the status of palæontology in these two fields."

1912h. New dinosaurs for the American Museum. Am. Mus. Jour. **12** (**6**): 219. O 1912.
[From the Cretaceous of Red Deer River, Canada.]

1912i. Facts and theories relating to the ancestry of man. Am. Mus. Jour. **12** (**7**): 255-256. N 1912.
[Critical mention of recent work of Schlosser, Boule, Elliot Smith, W. K. Gregory, Hrdlička, G. F. Wright, and others.]

1912j. News notes. Fossil walrus skull from Penobscot Bay. Am. Mus. Jour. **12** (**7**): 269. N 1912. [Unsigned.]
[*Trichechus rosmarus.*]

1912k. The ancestry of the edentates as illustrated by the skeleton of *Hapalops*, a Tertiary ancestor of the ground sloths. Am. Mus. Jour. **12** (**8**): 300-303. *2 figs.* D 1912.
[Reprinted with additions by the American Museum in 1913; pp. 1-8. *4 figs.*]
[Includes an illustrated phylogeny of the edentates.]

1912l. [Review of] "The Origin and Antiquity of Man." By G. F. Wright. Current Anthropological Literature **1** (**4**): 267-272. O-D 1912.

1913a. News notes. A tiny fossil skull. Am. Mus. Jour. **12** (**1**): 48. Ja 1913. [Unsigned.]
[Insectivore *Palæoryctes* from the Torrejon formation. Basal Eocene of New Mexico. See 1913i.]

1913b. The evolution of the horse in nature. [Revised and extended edition. With S. H. Chubb.] Am. Mus. Guide Leaflet Series **36**: 1-63. *38 figs.* Mr 1913.
"Illustrations mostly new, text revised."
[1st edition, 1903b; 4th edition, 1921a.]

1913c. Cuban fossil mammals; preliminary note. Bull. Geol. Soc. Am. **24**: 118-119. 24 Mr 1913.
[Regards land connections unnecessary for introduction of the Cuban mammalian fauna.]

1913d. News notes. [A visit to Rancho La Brea.] Am. Mus. Jour. **13** (**4**): 200. Ap 1913. [Unsigned.]

1913e. News notes. Dinosaur explorations in German East Africa. Am. Mus. Jour. **13** (**4**): 200. Ap 1913. [Unsigned.]

1913f. The laws of nomenclature in paleontology. Science **37** (**960**): 788-792. 23 My 1913.
[Discusses the "special difficulties that beset the vertebrate paleontologist." See 1913h.]

1913g. Certain theoretical considerations affecting phylogeny and correlation. Bull. Geol. Soc. Am. **24**: 283-292. *1 fig.* 1 Je 1913.
[Abstract, page 118.]
[Theories of dispersal, hybridization during migration and effects of remoteness from dispersal centers on diversity of type. Draws parallels from dispersal of human races.]

1913h. Nomenclature in paleontology. Science **38** (**968**): 87-88. 18 Jl 1913.
[Continuation of arguments presented in 1913f.]

1913i. A zalambdodont insectivore from the Basal Eocene. Bull. Am. Mus. Nat. Hist. **32**: 307-314. *6 figs. pls. 60, 61.* 25 Jl 1913.
[Abstract in Ann. N. Y. Acad. Sci. **23**: 263-264. *Palæoryctes puercensis* n. gen. & sp.]

1913j. American Museum expeditions for fossil vertebrates. Am. Mus. Jour. **13** (**6**): 286-287. O 1913.
[Cretaceous dinosaurs, Puerco mammals and work at Agate Springs, Nebraska.]

1913k. The asphalt group of fossil skeletons. The tar-pits of Rancho La Brea, California. Am. Mus. Jour. **13** (**7**): 290-297. *6 figs.* N 1913.

1913l. News notes. [Exhibit of fossil rhinoceroses.] Am. Mus. Jour. **13** (**8**): 374. 1913. [Unsigned.]

1914a. News notes. [Fossils from Rancho La Brea.] Am. Mus. Jour. **14** (**1**): 46-47. Ja 1914. [Unsigned.]

1914b. News notes. [New South American Fossils.] Am. Mus. Jour. **14** (**1**): 47-48. 1914. [Unsigned.]

1914c. Extinct vertebrates. [Report of the] Department of Vertebrate Palæontology. 45th Annual Report Am. Mus. Nat. Hist. 1913: 57-59. 1914.

1914d. [Note on *Felis atrox bebbi* of Rancho La Brea.] Nature **92**: 640. 5 F 1914.
["Appears to be as nearly related to the tiger as to the lion."]

1914e. Origin of Argentine wild horses. Nature **92**: 661. 12 F 1914.

1914f. Discussion [of C. Stock's paper on "The systematic position of the mylodont sloths from Rancho La Brea."] Bull. Geol. Soc. Am. **25**: 144. 30 Mr 1914.

1914g. Report of progress in the revision of the Lower Eocene faunas. (Abstract) Bull. Geol. Soc. Am. **25**: 144-145. 30 Mr 1914.

1914h. Introductory note to Albert Johannsen "Petrographic Analysis of the Bridger, Washakie, and other Eocene formations of the Rocky Mountains." Bull. Am. Mus. Nat. Hist. **33**: 209-210. 31 Mr 1914.

1914i. News notes. [Two new fossil mammals of the Lower Eocene.] Am. Mus. Jour. **14** (**4**): 167-168. Ap 1914. [Unsigned.] [See 1915n.]

1914j. Notes on Cuban fossil mammals. (Abstract) Ann. N. Y. Acad. Sci. **23**: 263-264. 31 Ap 1914.

1914k. News notes. [Expeditions for fossil vertebrates.] Am. Mus. Jour. **14** (**5**): 214. My 1914. [Unsigned.]

1914l. [Notes on auditory ossicles of rodents.] In "The auditory ossicles of American rodents." By T. D. A. Cockerell, and others. Bull. Am. Mus. Nat. Hist. **33**: 350-351, 379-380. 14 Jl 1914.

1914m. Time ratios in the evolution of mammalian phyla. A contribution to the problem of the age of the earth. Science **40** (**1024**): 232-235. 14 Au 1914.
[Assumption of a "fairly constant maximum rate of progressive evolution" would give a yard-stick for measurement of geologic time.]

1914n. Evidence of the Paleocene vertebrate fauna on the Cretaceous-Tertiary problem. Bull. Geol. Soc. Am. **25**: 381-402. *3 figs.* 15 S 1914.

"Characters of the Paleocene faunæ (Puerco, Torrejon, etc.), comparison with Cretaceous faunæ and with true Eocene faunæ, correlation and interpretation. Faunal migrations are due to widespread diastrophic movements accompanied by climatic changes, and constitute the best practical evidence of such widespread movements. The most marked faunal break is between the Paleocene and Eocene; between the Lance and Belly River there is a very considerable time interval, but no such marked break as should indicate a change in period. Between Lance and Paleocene there is an apparent faunal break but it is of doubtful interpretation. But until these doubtful points are cleared up it seems better to retain the division at this point which is the classical and generally accepted line between Cretaceous and Tertiary."

[See 1921d, 1921i.]

1914o. News notes. [Discovery of *Moropus* skeletons.] Am. Mus. Jour. **14** (**6-7**): 269. O-N 1914. [Unsigned.]

1914p. The largest known dinosaur. A huge extinct reptile from German East Africa, the largest known quadruped. Scientific American **111** (**22**): 443, 446-447. *1 fig.* 28 N 1914.

[*Gigantosaurus.*]

1914q. Note on *Felis atrox*. Nature **92**: 640. 1914.

1914r. New discoveries in the American Eocene. Report of the 83rd meeting, Brit. Assoc. Adv. Sci. 1913: 491. 1914.

[Differentiates the zones and faunas. Speaks of advances in studies on phylogenies and relationships of Tertiary Mammalia.]

1915a. The Tertiary sedimentary record and its problems. In "Problems of American Geology." By Charles Schuchert, and others. Yale University Press. Chapter 7: 377-478. *40 figs. map.* Ja 1915.

"Discussion of the Tertiary sedimentary formations of the Cordilleran region, their character and origin. Review of fossil record as contained therein showing the evolution of the Tertiary faunas."

1915b. News notes. [New exhibits of fossil vertebrates.] Am. Mus. Jour. **15** (**1**): 32. Ja 1915. [Unsigned.]

1915c. News notes. [New exhibits of fossil mammals.] Am. Mus. Jour. **15** (**2**): 86. F 1915. [Unsigned.]

1915d. Climate and Evolution. Ann. N. Y. Acad. Sci. **24**: 171-318. *33 figs.* 18 F 1915.

[The author's own copy of this famous work, in the Matthew Library, is extensively annotated.]

1915e. **————; & Granger, Walter.**

A revision of the Lower Eocene Wasatch and Wind River faunas. Part I. Order Feræ (Carnivora) Suborder Creodonta. By W. D. Matthew. Bull. Am. Mus. Nat. Hist. **34**: 1-103. *87 figs.* 3 Mr 1915.

[For continuation see, Matthew 1915l, 1915n, and 1918k.]

1915f. **de la Torre, Carlos; & ————**

Megalocnus and other Cuban ground sloths. (Abstract) Bull. Geol. Soc. Am. **26**: 152. 31 Mr 1915.

[*Megalocnus, Mesocnus, Miocnus,* and *Microcnus* from the Cuban Pleistocene. Nearest continental ally is *Megalonyx*. See 1931c.]

1915g. Affinities of *Hyopsodus*. (Abstract) Bull. Geol. Soc. Am. **26**: 152. 31 Mr 1915.

[Referred to the Condylarthra.]

1915h. Reconstruction of the skeleton of *Brachiosaurus*. (Abstract) Bull. Geol. Soc. Am. **20**: 153. 31 Mr 1915.

[From German East Africa. The largest known dinosaur.]

1915i. Ground-sloth from a cave in Patagonia. Am. Mus. Jour. **15** (**5**): 256. My 1915.

[*Grypotherium* as a contemporary of man.]

1915j. News notes. [Richard Lydekker, obituary notice.] Am. Mus. Jour. 15 (5): 263. My 1915. [Unsigned.]

1915k. Some remarkable extinct animals of South America. (Abstract) Ann. N. Y. Acad. Sci. **24**: 355. 14 My 1915.

1915l. ——————; **& Granger, Walter.**

A revision of the Lower Eocene Wasatch and Wind River Faunas. Part II. Order Condylarthra, Family Hyopsodontidæ. By W. D. Matthew. Bull. Am. Mus. Nat. Hist. **34**: 311-328. *10 figs.* 29 My 1915.

1915m. Dinosaurs. With special reference to the American Museum collections. Am. Mus. Handbook Series **5**: 1-124; 160-162. *48 figs. 1 pl.*

"A semi-popular review of the various kinds of dinosaurs, their antiquity, habits and environment, based chiefly upon American Museum collections."

[Contains chapters by S. W. Williston, H. F. Osborn and Barnum Brown.]

1915n. ——————; **& Granger, Walter.**

A revision of the Lower Eocene Wasatch and Wind River faunas. Part IV. Entelonychia, Primates, Insectivora (Part). Bull. Am. Mus. Nat. Hist. **34**: 429-483. *52 figs. pl. 15.* 24 S 1915.

"This revision is based upon the extensive collections and exact stratigraphic work of the American Museum expeditions of 1905, 1909-1916. Numerous new or little known mammals are described, and the affinities of all the Lower Eocene genera and species discussed; the phyletic and faunal changes and the successive levels in the Lower Eocene formations and their correlation fully shown. Part III by Granger covers the remaining Condylarthra; Part V, is listed under the year 1918 [1918]; Part VI, Tillodonta and Tæniodonta by Granger; Part VII, amblypods by Granger; Part VIII, Perissodactyla by Granger, and Part IX, conclusions, tabulation and correlation of faunas by Matthew and Granger not yet published."

1915o. News notes. [The Fulton mastodon.] Am. Mus. Jour. **15** (7): 373. N 1915. [Unsigned.]

1915p. Mammoths and mastodons. A guide to the collection of fossil proboscideans in the American Museum of Natural History. Am. Mus. Guide Leaflet Series **43**: 1-26. *11 figs. 1 pl.* N 1915.

[Author's annotated copy in the Matthew Library.]

1915q. ——————; **& Barnum Brown.**

Corythosaurus, the new duck-billed dinosaur. Am. Mus. Jour. **15** (**8**): 427-428. D 1915.

1915r. News notes. [Fossil Sirenian from Porto Rico.] Am. Mus. Jour. **15** (**8**): 432. D 1915. [Unsigned.]

1915s. **Cope, Edward Drinker; & ——————**

Hitherto unpublished plates of Tertiary Mammalia and Permian Vertebrata. Am. Mus. Nat. Hist. Monogr. Ser. **2**: Unpaged series of plates, with explanations. 1915.

[Published by the American Museum from plates in the custody of the U. S. Geol. Survey. The plates were prepared under the direction of E. D. Cope for the U. S. G. S. The descriptions of plates were written by Matthew.]

1915t. New discoveries in the Lower Eocene mammals. (Abstract) Ann. N. Y. Acad. Sci. **24**: 383. 14 My 1915.

1915u. ——————; **& Clarke, John M.**

Peccaries of the Pleistocene of New York. Bull. Geol. Soc. Am. **26**: 150-151. 31 Mr 1915.

1915v. [General consideration of paleontologic criteria in determining time relations.] (Discussion.) Bull. Geol. Soc. Am. **26**: 411. 23 N 1915.

[Problem of correlation by use of vertebrates. By title.]

1915w. [Report on fossil vertebrates of eastern Texas.] In "Problem of the Texas Tertiary sands." By E. T. Dumble. Bull. Geol. Soc. Am. **26**: 470-472. 4 D 1915.

[See 1920m.]

1915x. **Osborn, Henry Fairfield; & ———**
Exploration and researches of the Department of Vertebrate Palæontology. Preface to "Fossil vertebrates in the American Museum of Natural History." **4**: iii-xvi. 1909-1912. Ap 1915.

1916a. **Eastman, Charles R.; William King Gregory; & ———**
Record of progress in vertebrate palæontology. Science **43** (1099): 103-110. 21 Ja 1916.

1916b. In "Some remarks on Matthew's 'Climate and Evolution,' with Supplemental Note by W. D. Matthew." By Thomas Barbour. Ann. N. Y. Acad. Sci. **27**: 11-15. 26 Ja 1916.
[Discusses the origin of insular mammals.]

1916c. A new sirenian from the Tertiary of Porto Rico, West Indies. Ann. N. Y. Acad. Sci. **27**: 23-29. *2 figs.* 28 Ja 1916.
[Abstract, *ibid.* **26**: 439. 1915.]
[? *Halitherium antillense*, n. sp.]

1916d. The horse and his progenitors. Science Conspectus [Boston] **6** (1): 1-15. *9 figs. 1 pl.*

1916e. The grim wolf of the tar-pits. Am. Mus. Jour. **16** (1): 45-47. *2 figs.* Ja 1916.
"The great extinct wolf from the asphalt deposits at Rancho La Brea near Los Angeles."

1916f. News notes. [Dinosaurs from Alberta.] Am. Mus. Jour. **16** (1): 74-75. Ja 1916.

1916g. News notes. [Fossil reptile exhibits, *Myotragus,* Antillean fossils.] Am. Mus. Jour. **16** (2): 139. F 1916. [Unsigned.]

1916h. News notes. [Charles Falkenbach, obituary notice.] Am. Mus. Jour. **16** (3): 209. Mr 1916. [Unsigned.]

1916i. News notes. [Expeditions for fossil vertebrates.] Am. Mus. Jour. **16** (3): 209-211. Mr 1916. [Unsigned.]

1916j. A reptilian aëronaut. A new skeleton of *Pteranodon,* the giant flying reptile of the Cretaceous period. Am. Mus. Jour. **16** (4): 251-252. *1 fig.* Ap 1916.

1916k. The origin of the Pacific island faunas. Science **43** (1115): 686. 12 My 1916.

1916l. A marsupial from the Belly River Cretaceous. With critical observations upon the affinities of the Cretaceous mammals. Bull. Am. Mus. Nat. Hist. **35**: 477-500. *4 figs. pls. 2-6.* 24 Jl 1916.
"Description of a new genus *Eodelphis,* related to the opossums, redescription of *Thlaeodon padanicus* Cope and critical estimate of the remaining Cretaceous mammals with remarks upon the interpretation of the marsupial dentition."

1916m. Methods of correlation by fossil vertebrates. Bull. Geol. Soc. Am. **27**: 515-524. 1 S 1916.
"Discussion of principles of correlation. The Equidæ are regarded as affording the best standard for Tertiary correlation."

1916n. News notes. [Expeditions for fossil vertebrates.] Am. Mus. Jour. **16** (6): 412. O 1916. [Unsigned.]

1916o. Scourge of the Santa Monica Mountains [California]. Am. Mus. Jour. **16** (7): 469-472. *2 figs.* N 1916.
[The sabre-tooth cat, *Smilodon.*]

1916p. Kunz on Ivory and the Elephant. Am. Mus. Jour. **16** (8): 485-495. *8 pls.*

1916q. **Osborn, Henry Fairfield; & ———**
Explorations and researches of the Department of Vertebrate Palæontology. Preface to "Fossil Vertebrates in the American Museum of Natural History **5**: iii-viii. 1913-1914." 1916.

1917a. News notes. [Fossil tree trunks from the Cretaceous of Alberta.] Am. Mus. Jour. **17** (1): 78. Ja 1917. [Unsigned.]

1917b. News notes. [A Pliocene mastodon skeleton from Texas.] Am. Mus. Jour. **17** (2): 149. F 1917. [Unsigned.]

1917c. **Eastman, Charles R.; William King Gregory; & ———**
Recent progress in paleontology. Science 45 (1153): 117-121. 2 F 1917.

1917d. A fossil deer from Argentina, with a discussion of the distribution of various types of deer in North and South America. Am. Mus. Jour. 17 (3): 207-211. *2 figs.* Mr 1917.
"Notice of mounted skeleton of *Brachyceros pampæus,* with discussion of its affinities and the origin of the South American deer."

1917e. Gigantic *Megatherium* from Florida. (Abstract) Bull. Geol. Soc. Am. **28**: 212. 31 Mr 1917.
[This specimen "must have equaled or exceeded in bulk any known land mammal, living or extinct,"—with exception of *Baluchitherium,* which was not known at this date. Complete paper in Matthew, 1917, M. S.]

1917f. **———; & Granger, Walter.**
The skeleton of *Diatryma,* a gigantic bird from the Lower Eocene of Wyoming. Bull. Am. Mus. Nat. Hist. 37: 307-326. *1 fig. pls. 20-33.* 28 My 1917.
[Abstracts in Bull. Geol. Soc. Am. **28**: 212; and Science N.S. **46**: 246.]
"Description of a nearly complete skeleton of a gigantic flightless bird from the Wasatch formation of the Bighorn Basin. It has a huge skull and enormous compressed beak like the extinct South American *Phororhachos,* but it is not related to that genus, nor nearly related to any other known type. Its affinities are difficult to estimate; although having some features in common with the ostriches and cassowaries it does not appear to be related to them, but to be a very ancient terrestrial adaptation derived from primitive neognathine ('carinate') stock. Suggestions as to its bearing upon the general problem of the evolutionary history of birds."

1917g. Prehistoric animal life. The Mentor, Dept. Nat. Hist. 5 (13): 1-23. *14 figs. 6 pls.* 15 Au 1917.

1917h. A Paleocene bat. Bull. Am. Mus. Nat. Hist. 37: 569-571. *1 fig.* 7 S 1917.
[*Zanycteris paleocenus* n. gen. & sp. from base of the Wasatch, near Ignacio, Colorado.]

1917i. Absence of the pollex in Perissodactyla. Bull. Am. Mus. Nat. Hist. 37: 573-577. 7 S 1917.

1917j. Man and the Anthropoid. Science 46 (1184): 239-240. 7 S 1917.

1917k. The dentition of *Nothodectes.* Bull. Am. Mus. Nat. Hist. 37: 831-839. *pls. 99-102.* 5 D 1917.
"*Nothodectes* comes from a new fossil horizon at the top of the Paleocene, and is very closely allied to *Plesiadapis* of the Cernaysian of France. It is related both to Primates and Insectivora, especially to the Menotyphla."

1917l. A giant Eocene bird. Am. Mus. Jour. 17 (6): 417-418. *2 figs.* [See 1917f.]

1917m. News notes. [Duck-billed dinosaurs—Census of fossil skeletons.] Am. Mus. Jour. 17 (6): 419. O 1917. [Unsigned.]

1918a. **Gregory, William King; & ———**
Vertebrate palæontology. American Year Book. **1917**: 634-636. 1918.

1918b. The mounted skeleton of *Moropus* in the American Museum. A "clawed ungulate" from the middle Tertiary of Nebraska. Am. Mus. Jour. **18** (2): 120-123. *1 fig. 1 pl.* F 1918.

1918c. Generic nomenclature of the Proboscidea. (Abstract) Bull. Geol. Soc. Am. **29**: 141. 31 Mr 1918.

1918d. Affinities and phylogeny of the extinct Camelidæ. (Abstract) Bull. Geol. Soc. Am. **29**: 144. 31 M 1918.

1918e. **———; & Granger, Walter.**
Fossil mammals of the Tiffany Beds. (Abstract) Bull. Geol. Soc. Am. **29**: 152. 31 Mr 1918.
[Base of the Wasatch series in southern Colorado.]

1918f. Notes on the American Pliocene rhinoceroses. (Abstract) Bull. Geol. Soc. Am. **29**: 153. 31 Mr 1918.

1918g. Contributions to the Snake Creek fauna, with notes upon the Pleistocene of Western Nebraska, American Museum Expedition of 1916. Bull. Am. Mus. Nat. Hist. **38**: 183-229. *20 figs. pls. 4-10.* 18 Ap 1918.

"Fauna of the Snake Creek pockets recognized as including two distinct faunæ, Upper Miocene and Lower Pliocene, respectively. Various new types described. Revision of rhinoceros genera."

1918h. Skeletons of the Cuban ground sloth in the Havana and American Museums. Am. Mus. Jour. **18** (4): 312-313. *1 pl.* Ap 1918.

[*Megalocnus.* See 1931c.]

1918i. A fortunate collector. [William Stein, the discoverer of the giant bird, *Diatryma*, which was found in the Bighorn Basin of Wyoming.] Am. Mus. Jour. **18** (5): 389. *3 figs.* My 1918.

1918j. A Tertiary alligator. Am. Mus. Jour. **18** (6): 505-506. *1 fig.* O 1918.

1918k. ——————; **& Granger, Walter.**

A revision of the Lower Eocene Wasatch and Wind River Faunas. Part V. Insectivora (continued), Glires, Edentata. By W. D. Matthew. Bull. Am. Mus. Nat. Hist. **38**: 565-657. *68 figs.* 21 D 1918.

"Revision of Insectivora and Glires with description of various new or little known types. Description of a primitive genus of Metacheiromyidæ with discussion of its affinities. Regarded as a very primitive edentate related to early Insectivora and collateral ancestor of armadillos and other Xenarthra on one side, of Pholidota on the other side. Tubulidentata not related to true edentates." [See **1915e.**]

1918l. Affinities and origin of the Antillean mammals. Bull. Geol. Soc. Am. **29**: 657-666. [See 1919g and 1931c.]

1918m. News notes. [Fossils from Snake Creek beds.] Am. Mus. Jour. **18** (8): 731. D 1918. [Unsigned.]

1918n. **Osborn, Henry Fairfield; &** ——————

Explorations and researches of the Department of Vertebrate Palæontology. Preface to "Fossil Vertebrates in the American Museum of Natural History, **6**: iii-vii. 1915-1917." 1918.

1919a. ——————; **& Gregory, William King.**

Vertebrate palæontology. American Year Book, **1918**: 695-696. 1919.

1919b. Notes. [Lawrence M. Lambe, obituary notice.] Natural History **19** (3): 351-352. Mr 1919.

1919c. Reply to a letter from John Burroughs. Natural History **19** (4-5): 491-493. Ap-My 1919.

1919d. Honor to Adam Hermann. Natural History **19** (4-5): 491-493. Ap-My 1919.

"Address on occasion of the retirement of Mr. Hermann, head preparator in the American Museum department of vertebrate palæontology."

1919e. Notes. [The American Museum expedition to Jamaica.] Natural History **19** (6): 755. D 1919. [Unsigned.]

"Notice of the expedition of November, 1919, pointing out the evidence that the discovery of a fossil fauna will supply on palæogeography."

1919f. Notes. [Notice of expedition to the Agate fossil quarry.] Natural History **19** (6): 755. D 1919. [Unsigned.]

1919g. Recent discoveries of fossil vertebrates in the West Indies and their bearing on the origin of the Antillean fauna. Proc. Am. Phil. Soc. **58** (3): 161-181.

[Abstract in Science **49**: 546-547. See 1918l and 1931c.]

1920a. Plato's Atlantis in palæogeography. Proc. Nat. Acad. Sci. **6** (1): 17-18. Ja 1920.

[No scientific evidence for or against the existence of Atlantis.]

1920b. Flying reptiles. Natural History **20** (1): 73-81. *3 pl.* Ja-F 1920.

"Notice of the *Pteranodon* skeleton placed on exhibition, with some general remarks on pterodactyls."

1920c. ——————; **& Clarke, John M.**

Supposed fossil horse from the late Pleistocene found at Monroe, Orange Co., New York. (Abstract) Bull. Geol. Soc. Am. **31**: 204. Mr 1920.

[A native *Equus* survived the Wisconsin glaciation in eastern N. A.]

1920d. Status and limits of the Paleocene. (Abstract) Bull. Geol. Soc. Am. **31**: 221. 31 Mr 1920.

[Paleocene "includes the Puerco, Torrejon, Fort Union, and probably . . . the Cernaysian."]

1920e. New specimen of the Pleistocene bear *Arctotherium* from Texas. (Abstract) Bull. Geol. Soc. Am. **31**: 224-225. 31 Mr 1920.

1920f. John Campbell Merriam: New President of the Carnegie Institution. Natural History **20** (3): 253-254. *portrait*. My-Je 1920.

1920g. [Review.] "Die Stämme der Wirbeltiere." By Othenio Abel. Science **52** (**1332**): 37-38. 9 Jl 1920.

1920h. A new genus of rodents from the Middle Eocene. Jour. Mam. **1** (**4**): 168-169. Au 1920.

[*Reithroparamys* n. gen., type *Paramys delicatissimus* Leidy, from the Bridger.]

1920i. Social evolution: a palæontologist's viewpoint. Natural History **20** (**4**): 374-378. S-O 1920.

[A discussion of advances in social organization and their bearings on individual conduct.]

1920j. Three-toed horses. Natural History **20** (**4**): 473-478. *1 fig.* S-O 1920.

[Essay inspired by H. F. Osborn's monograph—"The Oligocene, Miocene, and Pliocene Equidae of North America."]

1920k. Canadian dinosaurs. Natural History **20** (**5**): 536-544. *3 figs.* *3 pls.* N-D 1920.

[*Ankylosaurus, Monoclonius, Deinodon, Corythosaurus, Struthiomimus, Saurolophus, Stephanosaurus, Prosaurolophus, Procheneosaurus, Kritosaurus, Trachodon, Gorgosaurus.*]

1920l. The proofs of the evolution of Man. Natural History **20** (**5**): 574-575. N-D 1920.

"Scientific evidence would indicate that the human race came from small tree-dwelling ancestors in Central Asia."

1920m. [Report on East Texas vertebrates.] Bull. Univ. Texas **1869**: 225, 231-233.

1921a. Evolution of the horse in nature. Am. Mus. Nat. Hist. Guide Leaflet Series **36** (**1**): 1-37. *25 figs.* *1 pl.* 4th ed. revised. Mr 1920.

[1st edition, 1903b.]

1921b. Notes on scientific museums of Europe. Natural History **21** (**2**): 185-190. Mr-Ap 1921.

1921c. A note on the Cernaysian mammal fauna. Amer. Jour. Sci. V. **1**: 509-511. Je 1921.

1921d. The Cannonball Lance formation. Science **54**: 27-29. Jl 1921.

[Vertebrate evidence summarized. (See 1914m, 1921i and 1922a.) On this basis the Lance is Cretaceous. By compromise with paleobotanists and stratigraphic geologists the Fort Union might be placed with it in the Cretaceous. The uppermost Paleocene faunas should be regarded as Tertiary.]

1921e. Notes. [Dr. Mook's researches on Crocodilia.] Natural History **21** (**4**): 433. Jl-Au 1921. [Unsigned.]

1921f. ——————; **& Granger, Walter.**

New genera of Paleocene mammals. Am. Mus. Novitates **13**: 1-7. 6 S 1921.

[*Ectypodus musculus* n. gen. & sp., *Eucosmodon* n. gen., *Paradectes elegans* n. gen. & sp., *Thylacodon pusillus* n. gen. & sp., *Leptacodon tener* n. gen. & sp., *Xenacodon mutilatus* n. gen. & sp., *Acmeodon secans* n. gen. & sp., *Labidolemur soricoides* n. gen. & sp., *Ignacius frugivorus* n. gen. & sp., *Navajovius kohlhaasæ* n. gen. & sp., *Carpodaptes aulacodon* n. gen. & sp., *Eoconodon* n. gen., *Mixoclænus encinensis* n. gen. & sp.]

1921g. *Stehlinius,* a new Eocene insectivore. Am. Mus. Novitates **14**: 1-5. 2 *figs.* 7 S 1921.
[*Stehlinius uintensis* n. gen. & sp.]

1921h. Life in other worlds. Science **54** (**1394**): 239-241. 16 S 1921.

1921i. Fossil vertebrates and the Cretaceous-Tertiary problem. Amer. Jour. Sci. V. **2**: 209-227. 1 O 1921.
[See 1921d. "Base of the true Eocene is the proper dividing line between Cretaceous and Tertiary."]

1921j. *Urus* and *Bison.* Natural History **21** (**6**): 598-606. *3 figs.* D 1921.

1921k. Why Palæontology? Natural History **21** (**6**): 639-641. *1 fig.* N-D 1921.

1921l. Notes. Palæontology [Jurassic of Cuba, Triassic of Texas, the ostrich dinosaur, and the Swabey mammoth skull.] Natural History **21** (**6**): 658-661. *2 figs.* N-D 1921.

1921m. Notes. [Snake Creek and Bolivian Mammals.] Natural History **21** (**6**): 661-662. N-D 1921. [Unsigned.]

1921n. A bibliographic list of the scientific publications of W. D. Matthew 1892-1921. Privately printed: Press of the New Era Printing Co. Lancaster, Pa. 31 D 1921.
[Accompanied by a mimeographed "List of Omissions."]

1922a. Phyletic relations of the Lance vertebrates. Pan-Am. Geol. **37** (**1**): 68-69. F 1922.
[Lower Paleocene should belong in Cretaceous; upper, in Tertiary. See 1921d and 1921i.]

1922b. "Errors and changes." In "The origin and evolution of the human dentition." By W. K. Gregory. Baltimore: Williams and Wilkins Co. xiii-xiv. 1922.
[Corrections offered by Matthew after examination of specimens in European collections and elsewhere.]

1922c. [Discussion of] "Stratigraphy of the lower Oreodon beds of the South Dakota big badlands." (Abstract) by W. J. Sinclair with discussion by W. D. Matthew. Bull. Geol. Soc. Am. **33**: 156. 31 Mr 1922.

1922d. New light on the phylogeny of the Canidæ. (Abstract) Bull. Geol. Soc. Am. **33**: 214. 31 Mr 1922.
[New material from the Lower Snake Creek is intermediate between Canidæ of the Oligocene and of the Pleistocene.]

1922e. Snake Creek fauna. (Abstract) Bull. Geol. Soc. Am. **33**: 215. 31 Mr 1922.

1922f. ——————; **& Brown, Barnum.**
The family Deinodontidae, with notice of a new genus from the Cretaceous of Alberta. Bull. Am. Mus. Nat. Hist. **46**: 367-385. *1 fig.* 31 My 1922.

1922g. A super-dreadnaught of the animal world, the armored dinosaur *Palæoscincus.* Natural History **22** (**4**): 333-342. *5 figs. 1 pl.* Jl-Au 1922.

1922h. ——————; **& Andrews, Roy Chapman.**
Gobi—a desert "wonder-house." Asia **22** (**12**): 859-862, 1000, 1002, and 1005. *8 figs.* D 1922.

1922i. **Osborn, Henry Fairfield; &** ——————
Explorations and researches of the Department of Vertebrate Palæontology. Preface to "Fossil Vertebrates in the American Museum of Natural History. **7**: iii-viii. 1918-1921." 1922.

1923a. Geological occurrence of the *Hesperopithecus* tooth. In "Notes on the type of *Hesperopithecus haroldcookii* Osborn." By W. K. Gregory and Milo Hellman. Am. Mus. Novitates **53**: 11-13. 6 Ja 1923.

1923b. ——————; **& Granger, Walter.**
Pliocene mammals of southern China. (Abstract) Bull. Geol. Soc. Am. **34**: 128. 30 Mr 1923.

1923c. Stratigraphy of the Snake Creek fossil quarries and the correlation of the faunas. (Abstract) Bull. Geol. Soc. Am. **34**: 131. 30 Mr 1923.

1923d. [Review.] "Grundzüge der Paläontologie. II Abt. Vertebrata." By K. A. von Zittel. Neuarbeitet von F. Broili und Max Schlosser. Science **58** (**1493**): 107-109. 10 Au 1923.

1923e. Scientific names of Greek derivation. Nature **112**: 241. 18 Au 1923.

1923f. Fossil Bones in the rock. The fossil quarry near Agate, Sioux County, Nebraska. Natural History **23** (**4**): 358-369. *9 figs. 2 pl.* Jl-Au 1923.

1923g. Recent progress and trends in vertebrate paleontology. (Presidential address to the Paleontological Society.) Bull. Geol. Soc. Am. **34** (**3**): 401-418. 30 S 1923. [Reprinted in Smithsonian Report for 1923, 273-289. 1925.]

1923h. Fossil vertebrates. Natural History **23** (**5**): 520-522. S-O 1923. [Unsigned.]

1923i. ——————; **& Brown, Barnum.**
Preliminary notices of skeletons and skulls of Deinodontidæ from the Cretaceous of Alberta. Am. Mus. Novitates **89**: 1-9. *5 figs.* 11 O 1923.

1923j. ——————; **& Granger, Walter.**
New fossil mammals from the Pliocene of Sze-Chuan, China. Bull. Am. Mus. Nat. Hist. **48**: 563-598. *27 figs.* 10 D 1923.
[An extensive fauna from cave deposits.]

1923k. ——————; **& Granger, Walter.**
The fauna of the Houldjin Gravels. Am. Mus. Novitates **97**: 1-6. *6 figs.* 18 D 1923.
[Oligocene of Mongolia.]

1923l. ——————; **& Granger, Walter.**
The fauna of the Ardyn Obo formation. Am. Mus. Novitates **98**: 1-5. 18 D 1923.
[Faunal list. *Ardynia præcox* n. gen. & sp.]

1923m. ——————; **& Granger, Walter.**
New Bathyergidæ from the Oligocene of Mongolia. Am. Mus. Novitates **101**: 1-5. *4 figs.* 28 D 1923.
[Rodents from Hsanda Gol formation. *Tsaganomys altaicus* n. gen. & sp. Tsaganomyinæ new subfamily; *Cyclomylus lohensis* n. gen. & sp.]

1923n. ——————; **& Granger, Walter.**
Nine new rodents from the [Hsanda Gol] Oligocene of Mongolia. Am. Mus. Novitates **102**: 1-10. *12 figs.* 31 D 1923.
[New genera: *Cricetops dormitor* n. gen. & sp., *Selenomys mimicus* n. gen. & sp., *Tataromys plicidens* n. gen. & sp., *Karakoromys decessus* n. gen. & sp., *Desmatolagus gobiensis* n. gen. & sp.]

1923o. Dinosaur remains of the Red Deer Valley [Alberta, Canada]. Illustrated folder. 4 pages.
[Issued by the Canadian Pacific Railway.]

1924a. Preliminary notes on the Mongolian faunas. Proc. Pan. Pac. Sci. Cong. (Melbourne, Australia, 1923) **1**: art. 66, sec. 7 (geology), div. 4 "Correlation of the Kainozoic formations of the Pacific region." 981-984. 1924.
[Faunal lists given. See 1926i.]

1924b. ——————; **& Granger, Walter.**
New Carnivora from the Tertiary of Mongolia. Am. Mus. Novitates **104**: 1-9. *7 figs.* 15 Ja 1924.
[Eocene and Oligocene; *Didymoconus colgatei* n. gen. & sp., *Amphicticeps shackelfordi* n. gen. & sp.]

1924c. ——————; **& Granger, Walter.**
New insectivores and ruminants from the Tertiary of Mongolia, with remarks on the correlation. Am. Mus. Novitates **105**: 1-7. *3 figs.* 18 Ja 1924.
[*Tupaiodon morrisi* n. gen. & sp., *Palæoscaptor acridens* n. gen. & sp., *Eumeryx culminis* n. gen. & sp. Includes a list of the Hsanda Gol, Oligocene, fauna, 28 mammals of which about one-half are new genera.]

1924d. Notes. Vertebrate Palæontology. Welcomed home by a dinosaur. Replicas of *Baluchitherium* distributed. Fossil Birds from Nebraska. Natural History **24** (1): 118-119. *1 fig.* Ja-F 1924.

1924e. Tertiary terrestrial vertebrate horizons of North America. Bull. Geol. Soc. Am. **35**: 172. 30 Mr 1924.
[Title only.]

1924f. Memorial to George F. Matthew. Bull. Geol. Soc. Am. **35**: 181-182. 30 Mr 1924.
[Summarized remarks.]

1924g. Exploration for fossil remains in Mongolia. (Abstract) Bull. Geol. Soc. Am. **35**: 187-188. 30 Mr 1924.

1924h. Third contribution to the Snake Creek fauna. Bull. Am. Mus. Nat. Hist. **50**: 59-210. *63 figs.* 3 Jl 1924.
[Stratigraphy of the Snake Creek, faunal lists, faunal zones, correlations; descriptions of mammals, phylogeny of the Canidæ, validity of species of Equidæ, distribution of Equidæ, antlers of *Merycodus, Chelydrops stricta* n. gen. & sp. of soft-shell turtle.]

1924i. Fossil animals of India. Natural History **24** (4): 208-214. *7 figs.* Jl-Au 1924.

1924j. A new link in the ancestry of the horse. Am. Mus. Novitates **131**: 1-2. 23 S 1924.
[Abstract in Brit. Assoc. Adv. Sci., Rept. 92nd meeting **1925**: 380-381.]
[*Plesippus* n. gen., type *Equus simplicidens* Cope. *Pliohippus proversus* Merriam, referred.]

1924k. Notes. Extinct animals. Fossil horses from the Texas Pliocene. Natural History **24** (5): 629-631. S-O 1924.

1924l. [Notes on *Serridentinus* and *"Baluchitherium."*] In *"Serridentiinus* and *Baluchitherium,* Loh formation, Mongolia." By H. F. Osborn. Am. Mus. Novitates **148**: 5. 11 N 1924.

1924m. Notes. Vertebrate fossils. [A note on collections made by F. von Huene in South America.] Natural History **24** (6): 726-727. N-D 1924.

1924n. Correlation of the Tertiary formations of the Great Plains. Bull. Geol. Soc. Am. **35**: 743-754. 30 D 1924.
[Criteria, value of the Equidæ and other phyla, characteristic genera, Eocene sequences, table.]

1925a. Notes. [Barnum Brown. Santiago Roth. Fossil mammals from St. Petersburg, Florida.] Natural History **25** (1): 96-97. Ja-F 1925.

1925b. Blanco and associated formations of northern Texas. (Abstract) Bull. Geol. Soc. Am. **36** (1): 221-222. 30 Mr 1925.

1925c. Fossil mammal faunas of Florida. (Abstract) Bull. Geol. Soc. Am. **36** (1): 225. 30 Mr 1925.

1925d. The value of palæontology. Natural History **25** (2): 166-168. Mr-Ap 1925.

1925e. ——————; **& Barbour, Erwin Hinckley.**
An American fossil giraffe, *Giraffa nebrascensis* n. sp. Bull. Neb. State Mus. **1** (4): 33-40. *figs. 14-20.* Ap 1925.
[Doubts are expressed as to the nature of this specimen. It consists of two worn cheek teeth.]

1925f. ——————; **& Granger, Walter.**
Fauna and correlation of the Gashato formation of Mongolia. Am. Mus. Novitates **189**: 1-12. *14 figs.* 7 O 1925.
[Age, somewhere between Lower Cretaceous and Upper Eocene. Notoungulate, multituberculate, rodent, creodont and (?) marsupial present. *Palæostylops iturus* n. gen. & sp., *Baënomys ambiguus* n. gen. & sp., *Prionessus lucifer* n. gen. & sp., *Eurymylus laticeps* n. gen. & sp., *Phenacolophus fallax* n. gen. & sp., *Hyracolestes ermineus* n. gen. & sp., *Sarcodon pygmæus* n. gen. & sp.]

1925g. ——————; **& Granger, Walter.**

New creodonts and rodents from the Ardyn Obo formation of Mongolia. Am. Mus. Novitates **193**: 1-7. *9 figs.* 27 O 1925.
[*Ardynictis furunculus* n. gen. & sp., *Ardynomys olseni* n. gen. & sp.]

1925h. ——————; **& Granger, Walter.**
New ungulates from the Ardyn Obo formation of Mongolia with faunal list and remarks on correlation. Am. Mus. Novitates **195**: 1-12. *13 figs.* 19 N 1925.
[*Paracolodon curtus* n. gen. & sp., *Miomeryx altaicus* n. gen. & sp.]

1925i. ——————; **& Granger, Walter.**
New mammals from the Shara Murun Eocene of Mongolia. Am. Mus. Novitates **196**: 1-11. *10 figs.* 20 N 1925.
[*Olsenia mira* n. gen. & sp., *Deperetella cristata* n. gen. & sp., *Cænolophus promissus* n. gen. & sp., *Archæomeryx aptatus* n. gen. & sp.—"an approximate ancestral type for the Pecora."]

1925j. ——————; **& Granger, Walter.**
New mammals from the Irdin Manha Eocene of Mongolia. Am. Mus. Novitates **198**: 1-12. *10 figs.* 21 N 1925.
[*Gobiohyus orientalis* n. gen. & sp.]

1925k. ——————; **& Granger, Walter.**
The smaller perissodactyls of the Irdin Manha formation, Eocene of Mongolia. Am. Mus. Novitates **199**: 1-9. *figs. 9.* 23 N 1925.
[Lophialetinae n. subfam., *Teleolophus medius* n. gen. & sp., *Lophialetes expeditus* n. gen. & sp.]

1925l. A dissenting opinion. In, "The origin of species. II Distinctions between rectigradations and allometrons." By H. F. Osborn. Proc. Nat. Acad. Sci. **11 (12)**: 751. D 1925.

1925m. ——————; **Madison, Harold Lester; and others.**
Code of ethics for museum workers. Being the report of a committee of the Am. Assoc. of Museums adopted unanimously at the twentieth meeting of the association 1925. The Am. Assoc. of Museums, New York **1925**: 8 pages. 1925.
[A preliminary draft of this code entitled "A tentative code of museum ethics" was drawn up by Harold L. Madison, Chairman of Committee on Ethics, and printed for discussion at the Twentieth Annual Meeting held at St. Louis, Missouri. May 17-21, 1925.]

1925n. **Osborn, Henry Fairfield; &** ——————
Explorations and researches of the Department of Vertebrate Palæontology. Preface to "Fossil Vertebrates in the American Museum of Natural History **8**: iii-x. 1922-1924." 1925.

1926a. Vertebrate Palæontology. American Year Book. **1925**: 933-934. 1926.

1926b. ——————; **& Granger, Walter.**
Two new perissodactyls from the Arshanto Eocene of Mongolia. Am. Mus. Novitates **208**: 1-5. *5 figs.* 16 F 1926.
[*Schlosseria magister* n. gen. & sp., *Teilhardia pretiosa* n. gen. & sp.]

1926c. [Note on rhinoceros remains from the Ricardo.] In "New canid and rhinocerotid remains from the Ricardo Pliocene of the Mohave Desert, California." By Chester Stock & Eustace L. Furlong. Univ. Calif. Publ. Geol. **16 (2)**: 50. 16 Mr 1926.

1926d. **Osborn, Henry Fairfield; &** ——————
Summary of discoveries in vertebrate paleontology in three seasons of field explorations in Mongolia. (1922, 1923, and 1925.) (Abstract) Bull. Geol. Soc. Am. **37**: 157-158. 30 Mr 1926. [See also 228.]
[Discovery of new Cretaceous and Tertiary faunas.]

1926e. ——————; **Holmes, W. W. & Megathlin, G. R.**
Pleistocene fauna from Seminole, Pinellas County, Florida. (Title only.) Bull. Geol. Soc. Am. **37**: 244. 30 Mr 1926.

1926f. The evolution of the horse. A record and its interpretation. Quart. Rev. Biol. 1 (2): 139-185. *27 figs. 1 pl.* Ap 1926.
[Critical summary of the evidence with many new facts and interpretations. Reduction of the phalanges, changes in carpus and tarsus, lower limbs, skull, and teeth. Morphological stages and the stratigraphic sequence in North America and Europe. Tables showing geologic and geographic distribution, and phylogeny of Perissodactyla. Evidences of evolution. Bibliography.]

1926g. On a new primitive deer and two traguloid genera from the Lower Miocene of Nebraska. Am. Mus. Novitates **215**: 1-8. *3 figs.* 10 My 1926.
[*Machæromeryx tragulus* n. gen. & sp.]

1926h. Early days of fossil hunting in the High Plains. Natural History **26** (5): 449-454. *3 pls.* S-O 1926.

1926i. Some comments on the characters of the fossil vertebrate faunas of Mongolia. (Abstract) Bull. Geol. Soc. China **5** (1): 5-9. 1926.
[Discusses positions of three Cretaceous and five Tertiary faunas. Abstract by Matthew in Biol. Abstracts **2** (3-5): abstract (7160). Mr-My 1928.]

1926j. The most significant fossil finds of the Mongolian expedition. In "Important results of the Central Asiatic expeditions." By C. P. Berkey, and others. Natural History **26** (5): 532-534. *5 figs. 9 pls.* S-O 1926.

1926k. Relations of vertebrates to sediments and sedimentary environments. In "Treatise on Sedimentation." By W. H. Twenhofel. 1st ed. 146-150.
[Also in 2nd ed. 183-186.]

1927a. Notes. The Siwalik beds. Natural History **27** (2): 187. Mr-Ap 1927.

1927b. [Various items] in "Geology of Mongolia." By C. P. Berkey and F. K. Morris. Nat. Hist. of Central Asia; Amer. Mus. Nat. Hist. **2**: i-xxxi + 1-475. *161 figs. 44 pls. 6 maps.*
["To Dr. W. D. Matthew, whose wide knowledge of palæontology and Cenozoic history make him an authority, we have referred many questions particularly as to the correlation of a number of the Tertiary formations." p. vii.]

1927c. A ballad of *Pithecanthropus.*
[Privately printed: 4. New York.]

1928a. The evolution of the mammals in the Eocene. Proc. Zool. Soc. London **1927**: 947-985. *16 figs.* 12 Ja 1928.
[Progressive differentiation from Triassic to Tertiary; Cretaceous placentals; Cretaceous character of the Paleocene mammals; conspectus of Paleocene and Lower Eocene forms; evolution through the Eocene; phylogeny of Carnivora, Perissodactyla, Artiodactyla. Theory of arboreal Eocene types, becoming diversified and specialized with development of brain and marked features in their evolution. Abstract by L. S. Russell in Biol. Abstracts. (2968). Ja 1932.]

1928b. *Xenotherium* an edentate. Jour. Mam. **9** (1): 70-71. F 1928.
[Lower Oligocene of Montana.]

1928c. ———; **& Siegfriedt, J. C. F.**
New fossil mammals from the Fort Union of Montana. (Abstract) Bull. Geol. Soc. Am. **39**: 300. 30 Mr 1928.
[Finding of marsupials suggests North American origin for cænolestids. MS. of entire paper in Matthew library.]

1928d. Affinities of some new paleocene mammals from Fort Union beds. (Abstract) Pan-Am. Geol. **49** (4): 319. My 1928.
[Also in Bull. Geol. Soc. Am. **40**: 258. 30 Mr 1928. (Title only.)]
[MS. of entire paper in Matthew library.]

1928e. Correlación de las formaciones Cenozoicas en la Argentina.—Carta del Dr. William D. Matthew a Alfredo Castellanos. New York, 11 de Octobre de 1923. Tradduccion del ingles. [Being the "Appendix" to Castellanos "Notas criticas

sobre el puelchense de las sedimentos neogenos de la Argentina." Rev. Univ. Nac. Cordova **15** (5-6): 52-54.] Jl-Au 1928.
[Discusses ground sloth distribution and occurrences in N. & S. Am. Also *Glyptodon,* carnivores, and equids.]

1928f. Dinosaurs and rickets. Sci. News Letter 15 Au 1928.
[Criticism of article by Marshall in S. N. L. Au 1928.]

1928g. Synopsis lectures in Paleontology 1. Outline and general principles of the history of life. Univ. Calif. Syllabus Series **213**: 1-253. *21 figs.* Au 1928.
[For 2nd ed. see Matthew 1930h.]
[Used in elementary classes at the University of California and elsewhere. A masterpiece of simplified presentation. Many of the figures were drawn by Matthew.]

1928h. ———; **Granger, Walter; & Simpson, George Gaylord.**
Paleocene multituberculates from Mongolia. Am. Mus. Novitates **331**: 1-4. *3 figs.* 30 O 1928.
[*Sphenopsalis nobilis* n. gen. & sp.; affinities of *Prionessus* and *Sphenopsalis.*]

1928i. [Review.] "Die Saugethiere, Einführung in die Anatomie und Systematik der recenten und fossilen Mammalia." By Max Weber. (2nd edition. Edited by O. Abel.) Science **68** (1766): 429-430. 2 N 1928.

1928j. The ape-man of Java. Natural History **28** (6): 577-588. *7 figs.* N-D 1928.
[A popularized word-picture of the life and environment of Pleistocene man. Reprints were distributed to the elementary classes in Paleontology at the University of California in 1928.]

1929a. A new and remarkable hedgehog from the later Tertiary of Nevada. Univ. Calif. Publ. Geol. **18** (4): 93-102. *pls. 7, 8.* 29 Ja 1929. [*Metechinus nevadensis* n. gen. & sp.]

1929b. Critical observations upon Siwalik mammals (exclusive of Proboscidea). Bull. Am. Mus. Nat. Hist. **56**: 437-560. *55 figs.* 9 F 1929.
[Result of re-examination of type collections. Correlation of faunas with Europe and N. Amer.; comments on faunal lists; review of collections in Calcutta and British Museum; phylogeny of Ursidæ; revision of the Machærodontinæ; notes on Siwalik rhinoceroses; notes on the Pondaung fauna of Burma; critical observations on Siwalik perissodactyls in British Museum; classification of the Chalicotheriidæ; phylogeny of Giraffidæ; notes on Camelidæ, Hippopotamidæ and Rodentia. Abstract by Matthew in Biol. Abstracts (2967) Ja 1932.]

1929c. On the phylogeny of horses, dogs and cats. Science **69** (1793): 494-496. 10 My 1929.
[An answer to Austin Clark's theory of evolution. Abstract by Matthew in Biol. Abstracts (26276). O 1930.]

1929d. Preoccupied names. Jour. Mam. **10** (2): 171. My 1929.
[*Stehlinius* preocc. becomes *Stehlinella. Palæolestes* preocc. becomes *Prodiacodon.*]

1929e. Reclassification of the artiodactyl families. Bull. Geol. Soc. Am. **40**: 403-408. 30 Je 1929.
["The customary grouping of the families of Artiodactyla is unsatisfactory as regards the extinct forms of the older Tertiary. The following arrangement is suggested;" Palæodonta, new suborder (Dichobunidæ and Entelodontidæ), Hyodonta, new suborder (dicotyls, pigs and hippopotamuses) Ancodonta, new suborder (anthracotheres, anoplotheres, cænotheres and oreodonts), Tylopoda (xiphodonts and camels), Pecora (true ruminants). Abstract by Matthew in Biol. Abstracts (26277). O 1930.]

1929f. The phylum in zoology and paleontology. Science **70** (1806): 142-143. 9 Au 1929.
[Continuation of the discussion with Austin Clark. Phylum, as used in Paleon-

tology designates what Clark has termed "linear evolution involving a time element."]

1929g. ——————; **Granger, Walter; & Simpson, George Gaylord.**
Additions to the fauna of the Gashato formation of Mongolia. Am. Mus. Novitates **376**: 1-12. *11 figs.* 7 O 1929.
[Revised faunal list included. *Praolestes nanus* n. gen. & sp., *Pseudictops lophiodon* n. gen. & sp., Eurymylidæ n. fam., *Opisthopsalis vetus* n. gen. & sp., *Prodinoceras martyr* n. gen. & sp.]

1929h. [Paleontological articles for] Encyclopædia Britannica, 14th edition. London and New York. Encyclopædia Britannica, Ltd. 24 Vols. 1929.
[Amblypoda **1**: 740; Carnivora: extinct forms **4**: 900-901; Creodonta **6**: 667, 668; Equidæ **8**: 670-672; *Glyptodon* **10**: 449; horse: fossil horses **11**: 757; *Machærodus* **14**: 574; *Megatherium* **15**: 214-215; *Mylodon* **16**: 42; *Palæotherium* **17**: 112; *Phenacodus* **17**: 698; Toxodontia **22**: 337; Tylopoda **22**: 641-642.]

1929i. [Note on distribution of Pleistocene mammals.] In "The pre-Illinoian Pleistocene geology of Iowa." By G. F. Kay and E. T. Apfel. Iowa Geol. Survey Ann. Rept. **34**: [122 and 194].

1930a. Boundary between the Pliocene and the Pleistocene. (Abstract) Bull. Geol. Soc. Am. **41** (1): 155. 31 M 1930.
[By title. Discussed by Dr. B. L. Clark.]

1930b. We discover a mastodon. Standard Oil Bulletin (Standard Oil Company of Calif.) **18**: 12-14. *3 figs.* My 1930.
[*Pliomastodon vexillarius* n. sp. This is the original description. See 1930j.]

1930c. ——————; **& Stirton, Ruben Arthur.**
Osteology and affinities of *Borophagus*. Univ. Calif. Publ. Geol. **19** (7): 171-216. *2 figs. pls. 21-34.* 9 My 1930.
[First paper on the Hemphill faunas, Pliocene of Texas. Includes preliminary faunal list of the Coffee Ranch Quarry at Miami, and phylogeny of the hyaenoid Canidæ.]

1930d. The phylogeny of dogs. Jour. Mam. **2** (2): 117-138. *3 figs.* My 1930.
[Canis derived from *Tephrocyon*, *Cyon* from *Temnocyon*, *Procyon* from *Phlaocyon*, *Ursus* through *Arctotherium* from *Hemicyon*.]

1930e. Credit or responsibility in scientific research. Science **71** (**1852**): 662-663. 27 Je 1930.

1930f. Range and limitations of species as seen in fossil faunas. Bull. Geol. Soc. Am. **41**: 271-274. 30 Je 1930.
[Abstract in Bull. Geol. Soc. Am. **41** (1): 210-211. 30 Mr 1930.]
["We should not . . . expect to find two or more closely related species living together at the same time, within the same area, and with the same habits, causing their remains to be preserved together in the same quarry."]

1930g. The dispersal of land animals. Scientia **48**: 33-42. Jl 1930.
[Continues the thesis advanced in "Climate and Evolution." See 1915d.]

1930h. Synopsis of lectures in Paleontology 1. Outline and general principles of the history of life. Univ. Calif. Syllabus Series **230**: 1-276. *26 figs.* Au 1930.
[2nd edition of 1928g.]

1930i. The pattern of evolution: a criticism of Doctor Austin Clark's thesis. Scientific American **143**: 192-196. *5 figs.* S 1930.
[See Austin Clark, Scientific American, **143**: 104 Au 1930.]

1930j. A Pliocene mastodon skull from California, *Pliomastodon vexillarius* n. sp. Univ. Calif. Publ. Geol. **19** (16): 335-348. *2 figs. pls. 41-44.* 26 N 1930.
[For original description see Matthew 1930b.]

1930k. ——————; **& Stirton, R. A.**
Equidæ from the Pliocene of Texas. Univ. Calif. Publ. Geol. **19** (17): 349-396. *pls. 45-58.* 29 N 1930.

[Continuation of study of faunas of the Hemphill beds. *Calippus* n. subgen., type *Protohippus placidus* Leidy. Distinctive characters of later Tertiary horses summarized.]

1931a. Critical observation on the phylogeny of the rhinoceroses. Univ. Calif. Publ. Geol. **20** (**1**): 1-9. *2 figs.* 23 Ja 1931.

[*Dicerorhinus* of Sumatra most primitive of living forms. Presence or absence of lower tusks, presence or absence of nasal and frontal horns are characters used to separate the main phyla. Modern rhinos regarded as relicts of the Tertiary forms.]

[Abstract Bull. Geol. Soc. Am. **42**: 366-367.]

1931b. [Abstract.] "The evolution of Earth and Man." By G. A. Baitsell, and others. Yale Univ. Press 1929. Biol. Abstracts **5** (**10**): abstract (22669). O 1931.

1931c. Genera and new species of ground sloths from the Pleistocene of Cuba. Am. Mus. Novitates **511**: 1-5. 16 D 1931.

[With prefatory note by Walter Granger. See 1915f, 1918h, 1918l, and 1919g. Three new species of *Mesocnus* and *Miocnus* are described.]

1932a. A review of the rhinoceroses with a description of *Aphelops* material from the Pliocene of Texas. (A posthumous paper edited by R. A. Stirton.) Univ. Calif. Publ. Geol. **20** (**12**): 411-482. *12 figs. pls. 61-79.* 26 F 1932.

[Characters of *Aphelops, Peraceras, Teleoceras;* comparisons with *Chilotherium* and other Old World genera; discussion of principles of dispersal of phyla.]

1932b. New fossil mammals from the Snake Creek quarries. Am. Mus. Novitates **540**: 1-8. *7 figs.* 16 Je 1932.

1932c. A skeleton of *Merycoidodon gracilis* and its adaptive significance. Univ. Calif. Publ. Geol. **22** (**5**): 13-30. *pls. 2, 3.* 12 Jl 1932.

[Species of *Merycoidodon;* adaptive significance of skull, habits, environment and causes of extinction of oreodonts.]

1932d. [Correspondence on the Hemphill County Pliocene faunas of Texas.] In "The geology of Hemphill County, Texas." By L. C. Reed, and O. M. Longnecker, Jr. Univ. Texas Bull. **3231**: 69-70. 1932.

[Discusses the age of the beds near Canadian and the origin of the deposits.]

1933a. ————; **& Mook, Charles Craig.**

New fossil mammals from the Deep River beds of Montana. Am. Mus. Novitates **601**: 1-7. *2 figs.* 22 Mr 1933.

[Pt. I. Occurrence, by C. C. Mook. Pt. II. Descriptions, by William Diller Matthew. *Brachyerix macrotis* n. sp. hedgehog. *Sciurus angusticeps* n. sp. squirrel.]

1937a. Paleocene faunas of the San Juan Basin, New Mexico. Trans. Am. Philos. Soc. **30**: i-viii + 1-510. *85 figs. pls. 1-55.*

NATURAL SCIENCES IN AMERICA

An Arno Press Collection

Allen, J[oel] A[saph]. **The American Bisons,** Living and Extinct. 1876

Allen, Joel Asaph. **History of the North American Pinnipeds:** A Monograph of the Walruses, Sea-Lions, Sea-Bears and Seals of North America. 1880

American Natural History Studies: The Bairdian Period. 1974

American Ornithological Bibliography. 1974

Anker, Jean. **Bird Books and Bird Art.** 1938

Audubon, John James and John Bachman. **The Quadrupeds of North America.** Three vols. 1854

Baird, Spencer F[ullerton]. **Mammals of North America.** 1859

Baird, S[pencer] F[ullerton], T[homas] M. Brewer and R[obert] Ridgway. **A History of North American Birds:** Land Birds. Three vols., 1874

Baird, Spencer F[ullerton], John Cassin and George N. Lawrence. **The Birds of North America.** 1860. Two vols. in one.

Baird, S[pencer] F[ullerton], T[homas] M. Brewer, and R[obert] Ridgway. **The Water Birds of North America.** 1884. Two vols. in one.

Barton, Benjamin Smith. **Notes on the Animals of North America.** Edited, with an Introduction by Keir B. Sterling. 1792

Bendire, Charles [Emil]. **Life Histories of North American Birds** With Special Reference to Their Breeding Habits and Eggs. 1892/1895. Two vols. in one.

Bonaparte, Charles Lucian [Jules Laurent]. **American Ornithology:** Or The Natural History of Birds Inhabiting the United States, Not Given by Wilson. 1825/1828/1833. Four vols. in one.

Cameron, Jenks. **The Bureau of Biological Survey:** Its History, Activities, and Organization. 1929

Caton, John Dean. **The Antelope and Deer of America:** A Comprehensive Scientific Treatise Upon the Natural History, Including the Characteristics, Habits, Affinities, and Capacity for Domestication of the Antilocapra and Cervidae of North America. 1877

Contributions to American Systematics. 1974

Contributions to the Bibliographical Literature of American Mammals. 1974

Contributions to the History of American Natural History. 1974

Contributions to the History of American Ornithology. 1974

Cooper, J[ames] G[raham]. **Ornithology.** Volume I, Land Birds. 1870

Cope, E[dward] D[rinker]. **The Origin of the Fittest:** Essays on Evolution and **The Primary Factors of Organic Evolution.** 1887/1896. Two vols. in one.

Coues, Elliott. **Birds of the Colorado Valley.** 1878

Coues, Elliott. **Birds of the Northwest.** 1874

Coues, Elliott. **Key To North American Birds.** Two vols. 1903

Early Nineteenth-Century Studies and Surveys. 1974

Emmons, Ebenezer. **American Geology:** Containing a Statement of the Principles of the Science. 1855. Two vols in one.

Fauna Americana. 1825-1826

Fisher, A[lbert] K[enrick]. **The Hawks and Owls of the United States in Their Relation to Agriculture.** 1893

Godman, John D. **American Natural History:** Part I — Mastology and **Rambles of a Naturalist.** 1826-28/1833. Three vols. in one.

Gregory, William King. **Evolution Emerging:** A Survey of Changing Patterns from Primeval Life to Man. Two vols. 1951

Hay, Oliver Perry. **Bibliography and Catalogue of the Fossil Vertebrata of North America.** 1902

Heilprin, Angelo. **The Geographical and Geological Distribution of Animals.** 1887

Hitchcock, Edward. **A Report on the Sandstone of the Connecticut Valley,** Especially Its Fossil Footmarks. 1858

Hubbs, Carl L., editor. **Zoogeography.** 1958

[Kessel, Edward L., editor]. **A Century of Progress in the Natural Sciences: 1853-1953.** 1955

Leidy, Joseph. **The Extinct Mammalian Fauna of Dakota and Nebraska,** Including an Account of Some Allied Forms from Other Localities, Together with a Synopsis of the Mammalian Remains of North America. 1869

Lyon, Marcus Ward, Jr. **Mammals of Indiana.** 1936

Matthew, W[illiam] D[iller]. **Climate and Evolution.** 1915

Mayr, Ernst, editor. **The Species Problem.** 1957

Mearns, Edgar Alexander. **Mammals of the Mexican Boundary of the United States.** Part I: Families Didelphiidae to Muridae. 1907

Merriam, Clinton Hart. **The Mammals of the Adirondack Region,** Northeastern New York. 1884

Nuttall, Thomas. **A Manual of the Ornithology of the United States and of Canada.** Two vols. 1832-1834

Nuttall Ornithological Club. **Bulletin of the Nuttall Ornithological Club:** A Quarterly Journal of Ornithology. 1876-1883. Eight vols. in three.

[Pennant, Thomas]. **Arctic Zoology. 1784-1787.** Two vols. in one.

Richardson, John. **Fauna Boreali-Americana;** Or the Zoology of the Northern Parts of British America, Containing Descriptions of the Objects of Natural History Collected on the Late Northern Land Expeditions Under Command of Captain Sir John Franklin, R. N. Part I: Quadrupeds. 1829

Richardson, John and William Swainson. **Fauna Boreali-Americana:** Or the Zoology of the Northern Parts of British America, Containing Descriptions of the Objects of Natural History Collected by the Late Northern Land Expeditions Under Command of Captain Sir John Franklin, R. N. Part II: The Birds. 1831

Ridgway, Robert. **Ornithology.** 1877

Selected Works By Eighteenth-Century Naturalists and Travellers. 1974

Selected Works in Nineteenth-Century North American Paleontology. 1974

Selected Works of Clinton Hart Merriam. 1974

Selected Works of Joel Asaph Allen. 1974

Selections From the Literature of American Biogeography. 1974

Seton, Ernest Thompson. **Life-Histories of Northern Animals: An Account of the Mammals of Manitoba.** Two vols. 1909

Sterling, Keir Brooks. **Last of the Naturalists:** The Career of C. Hart Merriam. 1974

Vieillot, L. P. **Histoire Naturelle Des Oiseaux de L'Amerique Septentrionale,** Contenant Un Grand Nombre D'Especes Decrites ou Figurees Pour La Premiere Fois. 1807. Two vols. in one.

Wilson, Scott B., assisted by A. H. Evans. **Aves Hawaiienses:** The Birds of the Sandwich Islands. 1890-99

Wood, Casey A., editor. **An Introduction to the Literature of Vertebrate Zoology.** 1931

Zimmer, John Todd. **Catalogue of the Edward E. Ayer Ornithological Library.** 1926